Honda 100/125 Single cylinder models Owners Workshop Manual

by Clive Brotherwood
with an extra Chapter on the CB125J model (CB125S-76) by Jeff Clew

Models covered from 1970 onwards

CB100	99cc
CL100	99cc
SL100	99cc
CB125S	124cc
CD125S	124cc
SL125 Trail	124cc
CB125J	124cc (First introduced UK Oct 1975)

ISBN 978 0 85696 188 5

© J H Haynes & Co. Ltd. 1995

All rights reserved. No part of this book may be reproduced or transmitted in any form or by any means, electronic or mechanical, including photocopying, recording or by any information storage or retrieval system, without permission in writing from the copyright holder.

(188-3U7)

J H Haynes & Co. Ltd.
Haynes North America, Inc

www.haynes.com

Acknowledgements

Grateful thanks are due to Honda (UK) Limited for the technical assistance given whilst this manual was being prepared, and for permission to reproduce their drawings. Fran Ridewood and Co of Wells provided the CB 125S model which was used for the photographic sequences, and supplied the necessary spare parts used during the rebuild. Brian Horsfall assisted with the dismantling and rebuilding sequences and devised the various ingenious methods for overcoming the lack of service tools. Les Brazier arranged and took the photographs; Jeff Clew edited the text and Rod Grainger planned the layout of each page.

We should also like to acknowledge the help of the Avon Rubber Company who kindly advised on tyre fitting, NGK Spark Plugs (UK) Limited for the provision of spark plug photographs, and Vincent and Jerrom Limited, East Reach, Taunton, who kindly supplied the model used for the cover illustration.

About this manual

The author of this manual has the conviction that the only way in which a meaningful and easy to follow text can be written is first to do the work himself, under conditions similar to those found in the average household. As a result, the hands seen in the photographs are those of the author. Even the machine photographed was not new; an example that had covered several thousand miles was selected so that the conditions encountered would be similar to those found by the average rider. Unless specially mentioned, and therefore considered essential, Honda service tools have not been used. There is invariably some alternative means of slackening or removing some vital component when service tools are not available and risk of damage has to be avoided at all costs.

Each of the seven chapters is divided into numbered sections. Within the sections are numbered paragraphs. In consequence, cross reference throughout this manual is both straightforward and logical. When a reference is made 'See Section 5.12' it means Section 5, paragraph 12 in the same Chapter. If another Chapter were meant, the text would read 'See Chapter 2, Section 5.12'.

All photographs are captioned with a section/paragraph number to which they refer and are always relevant to the chapter text adjacent.

Figure numbers (usually line illustrations) appear in numerical order, within a given chapter. Fig 1.1 therefore refers to the first figure in Chapter 1. Left hand and right hand descriptions of the machines and their component parts refer to the right and left of a given machine when the rider is seated normally.

Motorcycle manufacturers continually make changes to specifications and recommendations, and these, when notified, are incorporated into our manuals at the earliest opportunity

We take great pride in the accuracy of information given in this manual, but motorcycle manufacturers make alterations and design changes during the production run of a particular motorcycle of which they do not inform us. No liability can be accepted by the authors or publishers for loss, damage or injury caused by any errors in, or omissions from, the information given.

Contents

Chapter	Section	Page
Introductory pages	Acknowledgements	2
	About this manual	2
	Introduction to the Honda 100/125 single cylinder machines	5
	Buying spare parts	5
	Routine maintenance	6
	Safety first!	8
Chapter 1/Engine, clutch and gearbox	Specifications	9, 10, 11
	Engine/gearbox removal	11
	Dismantling	14
	Examination and renovation	21
	Reassembly	29
	Fault diagnosis	39, 40
Chapter 2/Fuel system and lubrication	Specifications	41
	Carburettor	45
	Exhaust system	45
	Oil pump	47
	Fault diagnosis	48
Chapter 3/Ignition system	Specifications	49
	Contact breakers	49
	Spark plug	53
	Fault diagnosis	53
Chapter 4/Frame and forks	Specifications	54
	Front forks - removal, dismantling and examination	56
	Front forks - reassembly and refitting	58
	Frame	61
	Swinging arm	62
	Rear suspension units	62, 65
	Fault diagnosis	68
Chapter 5/Wheels, brakes and tyres	Specifications	69
	Wheels	72, 75
	Brakes	72, 75
	Final drive chain	75
	Tyres	76
	Fault diagnosis	78
Chapter 6/Electrical system	Specifications	79
	Alternator	79, 80
	Battery	80, 81
	Bulb replacement	81, 82
	Fuses	83
	Wiring diagrams	84 - 89
	Fault diagnosis	90
Chapter 7/The CB125J model (CB125S-76)		91
Conversion table		104
Index		105
English / American terminology		107

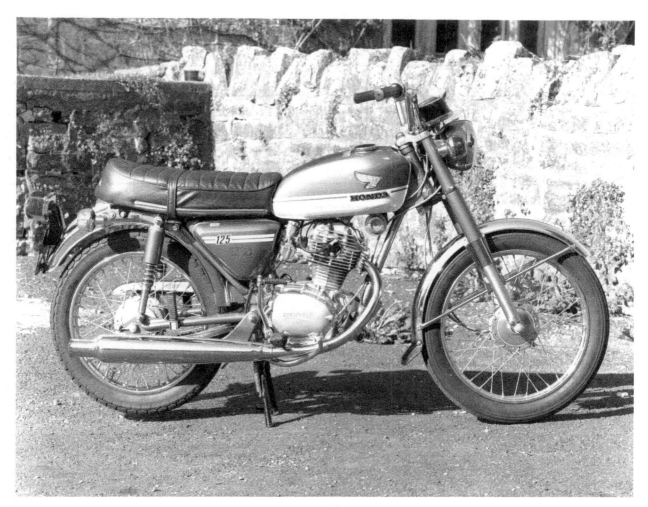

The Honda 125

Model dimensions

	CB100	CB125S	SL125
Overall length	74.2 in (1885 mm)	74.02 in (1880 mm)	78.5 in (1995 mm)
Overall width	29.5 in (750 mm)	29.5 in (750 mm)	31.9 in (810 mm)
Overall height	40.0 in (1015 mm)	40.0 in (1015 mm)	44.3 in (1115 mm)
Wheelbase	47.4 in (1205 mm)	40.0 in (1205 mm)	50.2 in (1275 mm)
Weight	198.4 lb (90 kg)	200.0 lb (91 kg)	209.5 lb (95.0 kg)

	CL 100	SL 100	CD 125S
Overall length	71.6 in (1820 mm)	75.4 in (1915 mm)	74.8 in (1995 mm)
Overall width	32.5 in (825 mm)	31.9 in (810 mm)	31.9 in (810 mm)
Overall height	40.5 in (1030 mm)	42.9 in (1090 mm)	39.4 in (1000 mm)
Wheelbase	49.4 in (1255 mm)	49.4 in (1255 mm)	47.2 in (1200 mm)
Weight	191.8 lb (87 kg)	211.7 lb (96 kg)	196.2 lb (89 kg)

Introduction to the Honda 100·125 c.c. single cylinder machine

The Honda 100 cc and 125 cc single cylinder machines have rapidly become very popular as small engined, spritely machines. Honda started with the 100 cc single in 1970 and developed it to the stage where they have incorporated the neat little engine unit in a very attractive 125 cc trail bike. No doubt Honda has not, by any means, fully exploited the original design, and with their customary ingenuity will continue to make the 125 cc unit even better.

Modifications to the Honda singles range

The Honda CB100 and CB125S singles, along with their variants, have changed but little since their introduction a few years ago. Most of the modifications made have affected what may best be termed the cosmetic appearance of each, using new colour schemes and different forms of tank styling. The two trail models were added to the range at a later stage, following the upsurge of interest in off-road riding. For these models a certain amount of redesign proved necessary in terms of the cycle parts and in consequence both models have a quite different frame and modified front and rear suspension. They also have different size wheels and tyres, to give the required increase in ground clearance and improved wheel grip for crossing rough terrain. Even so, there is still a strong family resemblance with the standard road versions, especially since the engine unit is virtually identical.

Buying spare parts

When ordering spare parts for any Honda it is advisable to deal direct with an official Honda agent who should be able to supply most of the parts ex-stock. Parts cannot be obtained from Honda (UK) Limited direct; all orders must be routed via an approved agent, even if the parts required are not held in stock.

Always quote the engine and frame numbers in full, particularly if parts are required for any of the earlier models. The frame number is stamped on the steering head reading downwards towards the base. The engine number is stamped in close proximity to the gearchange lever. Lastly, it is advisable to make note of the colour scheme, especially if any cycle parts are to be included in the order.

Use only parts of genuine Honda manufacture. Pattern parts are available, some of which originate from Japan and are packed in boxes of similar design to the manufacturer's originals. Pattern parts do not necessarily make a satisfactory replacement, even if there is an initial price advantage. Many cases are on record where reduced life or sudden failure has occurred, to the detriment of performance and reliability.

Some of the more expendable parts such as spark plugs, bulbs, tyres, oils and greases etc can be obtained from accessory shops and motor factors, who have convenient opening hours, charge lower prices and can often be found not far from home. It is also possible to obtain parts on a Mail Order basis from a number of specialists who advertise regularly in the motorcycle magazines.

 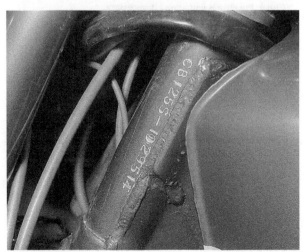

Orientation of engine and frame numbers

Routine maintenance

Periodic routine maintenance is a continuous process that commences immediately the machine is used. It must be carried out at specified mileage recordings or on a calendar date basis if the machine is not used regularly, whichever falls soonest. Maintenance should be regarded as an insurance policy, to help keep the machine in peak condition and to ensure long, trouble-free service. It has the additional benefit of giving early warning of any faults that may develop and will act as a regular safety check, to the obvious advantage of both rider and machine alike.

The various maintenance tasks are described under their respective mileage and calendar headings. Accompanying diagrams are provided where necessary. It should be remembered that the interval between the various maintenance tasks serves only as a guide. As the machine gets older or is used under particularly adverse conditions, it would be advisable to reduce the period between each check.

Some of the tasks are described in detail, where they are not mentioned fully as routine maintenance items in the text. If a specific item is mentioned, but not described in detail, it will be covered fully in the appropriate chapter. No special tools are required for the normal routine maintenance tasks. The tools contained in the tool kit supplied with every new machine will prove adequate for each task, or if they are not available, the tools found in the average household will usually suffice.

Check electrolyte level of battery at frequent intervals

Monthly or every 200 miles (300 km)

Change the engine oil (new machine) – this should then be changed after every 1000 miles (1600 km).
Clean the oil filter.
Check ignition timing and adjust if necessary.
Check valve tappet clearance and adjust if necessary.
Adjust the cam chain.
Check the clutch and adjust if necessary.
Adjust the drive chain and sprockets and lubricate.
Adjust the rear and front brakes.
Check the wheel rims and spokes.
Check the battery electrolyte level and replenish if necessary.

Three monthly or every 3000 miles (5000 km)

Clean, adjust or replace the spark plug.
Check or service the contact breaker points.
Check or adjust the ignition timing.
Check or adjust the valve tappet clearance.
Adjust the cam chain.
Clean the air cleaner.
Check the throttle operation.
Check or adjust the carburettor.
Clean the fuel tap gauze.
Check the fuel tank and fuel lines.
Check or adjust the clutch.
Adjust the drive chain and sprockets and lubricate.
Adjust the front and rear brakes.

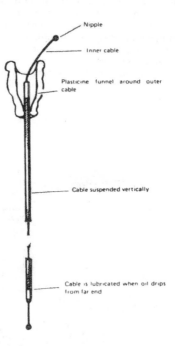

Oiling a control cable

Check the front and rear brake links.
Check the wheel rims and spokes.
Check or replace the tyres.
Check the front fork oil.
Check the side-stand spring.
Check the battery level and replenish if necessary.
Lights, horn and speedometer — check for operation or adjust.

Six monthly or every 6000 miles (10,000 km)

Oil filter — clean.
Spark plug — clean and adjust or replace.
Contact breaker points — check or service.
Ignition timing — check or adjust.
Valve tappet clearance — check or adjust.
Cam chain — adjust.
Throttle operation — check.
Carburettor — check or adjust.
Fuel tap gauze — clean.
Fuel tank and fuel lines — check.
Clutch — check or adjust.
Drive chain and sprockets — adjust and lubricate or replace.
Front and rear brake — adjust.
Front and rear brake shoes — check or replace.
Front and rear brake links — check.
Wheel rims and spokes — check.
Tyres — check or replace.
Change the front fork oil.
Steering head bearings — check or adjust.
Steering head lock — check for operation.
Side stand spring — check.
Battery level — replenish if necessary.
Lights, horn and speedometer — check for operation or adjust.

Yearly or every 12,000 miles (19,200 km)

Clean the oil filter.
Clean the air cleaner.
Check or replace the front and rear brake shoes.
Check and change the front fork oil.
Check or adjust the steering head bearings.
Check the steering head lock for operation.

Quick reference maintenance data

Engine oil	1.75 Imp gal (1.0 litre)
Fuel tank	
SL125	1.54 Imp gal (7.0 litres), inc. 0.33 Imp gal (1.5 litre) reserve
Other models	1.65 Imp gal (7.5 litres), inc. 0.26 Imp gal (1.2 litre) reserve
Valve tappet clearance	0.002 in (0.05 mm)
Contact breaker gap (points gap)	0.012 – 0.016 in (0.3 – 0.4 mm)
Spark plug gap	0.024 – 0.028 in (0.6 – 0.7 mm)
Tyre pressures	Front 26 psi Rear 28 psi

Recommended lubricants

Component	Castrol Product
Engine/gearbox	Castrol GTX
Front forks	Castrol TQF
Final drive chain	Castrol Graphited Grease
All greasing points	Castrol LM Grease

Safety first!

Professional motor mechanics are trained in safe working procedures. However enthusiastic you may be about getting on with the job in hand, do take the time to ensure that your safety is not put at risk. A moment's lack of attention can result in an accident, as can failure to observe certain elementary precautions.

There will always be new ways of having accidents, and the following points do not pretend to be a comprehensive list of all dangers; they are intended rather to make you aware of the risks and to encourage a safety-conscious approach to all work you carry out on your vehicle.

Essential DOs and DON'Ts

DON'T start the engine without first ascertaining that the transmission is in neutral.

DON'T suddenly remove the filler cap from a hot cooling system – cover it with a cloth and release the pressure gradually first, or you may get scalded by escaping coolant.

DON'T attempt to drain oil until you are sure it has cooled sufficiently to avoid scalding you.

DON'T grasp any part of the engine, exhaust or silencer without first ascertaining that it is sufficiently cool to avoid burning you.

DON'T allow brake fluid or antifreeze to contact the machine's paintwork or plastic components.

DON'T syphon toxic liquids such as fuel, brake fluid or antifreeze by mouth, or allow them to remain on your skin.

DON'T inhale dust – it may be injurious to health (see *Asbestos* heading).

DON'T allow any spilt oil or grease to remain on the floor – wipe it up straight away, before someone slips on it.

DON'T use ill-fitting spanners or other tools which may slip and cause injury.

DON'T attempt to lift a heavy component which may be beyond your capability – get assistance.

DON'T rush to finish a job, or take unverified short cuts.

DON'T allow children or animals in or around an unattended vehicle.

DON'T inflate a tyre to a pressure above the recommended maximum. Apart from overstressing the carcase and wheel rim, in extreme cases the tyre may blow off forcibly.

DO ensure that the machine is supported securely at all times. This is especially important when the machine is blocked up to aid wheel or fork removal.

DO take care when attempting to slacken a stubborn nut or bolt. It is generally better to pull on a spanner, rather than push, so that if slippage occurs you fall away from the machine rather than on to it.

DO wear eye protection when using power tools such as drill, sander, bench grinder etc.

DO use a barrier cream on your hands prior to undertaking dirty jobs – it will protect your skin from infection as well as making the dirt easier to remove afterwards; but make sure your hands aren't left slippery. Note that long-term contact with used engine oil can be a health hazard.

DO keep loose clothing (cuffs, tie etc) and long hair well out of the way of moving mechanical parts.

DO remove rings, wristwatch etc, before working on the vehicle – especially the electrical system.

DO keep your work area tidy – it is only too easy to fall over articles left lying around.

DO exercise caution when compressing springs for removal or installation. Ensure that the tension is applied and released in a controlled manner, using suitable tools which preclude the possibility of the spring escaping violently.

DO ensure that any lifting tackle used has a safe working load rating adequate for the job.

DO get someone to check periodically that all is well, when working alone on the vehicle.

DO carry out work in a logical sequence and check that everything is correctly assembled and tightened afterwards.

DO remember that your vehicle's safety affects that of yourself and others. If in doubt on any point, get specialist advice.

IF, in spite of following these precautions, you are unfortunate enough to injure yourself, seek medical attention as soon as possible.

Asbestos

Certain friction, insulating, sealing, and other products – such as brake linings, clutch linings, gaskets, etc – contain asbestos. *Extreme care must be taken to avoid inhalation of dust from such products since it is hazardous to health*. If in doubt, assume that they *do* contain asbestos.

Fire

Remember at all times that petrol (gasoline) is highly flammable. Never smoke, or have any kind of naked flame around, when working on the vehicle. But the risk does not end there – a spark caused by an electrical short-circuit, by two metal surfaces contacting each other, by careless use of tools, or even by static electricity built up in your body under certain conditions, can ignite petrol vapour, which in a confined space is highly explosive.

Always disconnect the battery earth (ground) terminal before working on any part of the fuel or electrical system, and never risk spilling fuel on to a hot engine or exhaust.

It is recommended that a fire extinguisher of a type suitable for fuel and electrical fires is kept handy in the garage or workplace at all times. Never try to extinguish a fuel or electrical fire with water.

Note: *Any reference to a 'torch' appearing in this manual should always be taken to mean a hand-held battery-operated electric lamp or flashlight. It does **not** mean a welding/gas torch or blowlamp.*

Fumes

Certain fumes are highly toxic and can quickly cause unconsciousness and even death if inhaled to any extent. Petrol (gasoline) vapour comes into this category, as do the vapours from certain solvents such as trichloroethylene. Any draining or pouring of such volatile fluids should be done in a well ventilated area.

When using cleaning fluids and solvents, read the instructions carefully. Never use materials from unmarked containers – they may give off poisonous vapours.

Never run the engine of a motor vehicle in an enclosed space such as a garage. Exhaust fumes contain carbon monoxide which is extremely poisonous; if you need to run the engine, always do so in the open air or at least have the rear of the vehicle outside the workplace.

The battery

Never cause a spark, or allow a naked light, near the vehicle's battery. It will normally be giving off a certain amount of hydrogen gas, which is highly explosive.

Always disconnect the battery earth (ground) terminal before working on the fuel or electrical systems.

If possible, loosen the filler plugs or cover when charging the battery from an external source. Do not charge at an excessive rate or the battery may burst.

Take care when topping up and when carrying the battery. The acid electrolyte, even when diluted, is very corrosive and should not be allowed to contact the eyes or skin.

If you ever need to prepare electrolyte yourself, always add the acid slowly to the water, and never the other way round. Protect against splashes by wearing rubber gloves and goggles.

Mains electricity and electrical equipment

When using an electric power tool, inspection light etc, always ensure that the appliance is correctly connected to its plug and that, where necessary, it is properly earthed (grounded). Do not use such appliances in damp conditions and, again, beware of creating a spark or applying excessive heat in the vicinity of fuel or fuel vapour. Also ensure that the appliances meet the relevant national safety standards.

Ignition HT voltage

A severe electric shock can result from touching certain parts of the ignition system, such as the HT leads, when the engine is running or being cranked, particularly if components are damp or the insulation is defective. Where an electronic ignition system is fitted, the HT voltage is much higher and could prove fatal.

Chapter 1 Engine, clutch and gearbox

Contents

General description ... 1	Gearchange selector mechanism - examination and renovation 22
Operations with the engine in the frame ... 2	Engine reassembly - general ... 23
Removing the engine/gearbox unit ... 3	Engine reassembly - replacing the crankshaft assembly ... 24
Dismantling the engine - general ... 4	Engine reassembly - replacing the gear cluster and selectors ... 25
Dismantling the engine - removing the cylinder head, barrel and piston ... 5	Engine reassembly - joining the crankcases ... 26
Dismantling the engine - removing the alternator, cam chain and final drive sprocket ... 6	Engine reassembly - replacing the clutch, gearchange mechanism, primary drive and oil pump ... 27
Dismantling the engine - removing the centrifugal oil filter, oil pump, clutch and gear selector mechanism ... 7	Engine reassembly - replacing the centrifugal oil filter ... 28
	Engine reassembly - replacing the right hand engine cover ... 29
Dismantling the engine - separating the crankcases, removing the crankshaft assembly and gear cluster ... 8	Engine reassembly - replacing the cam chain, tensioner and alternator ... 30
Examination and renovation - general ... 9	Engine reassembly - replacing the piston, cylinder barrel and cam chain guide ... 31
Main bearings and oil seals - examination and renovation ... 10	Engine reassembly - replacing the cylinder head and camshaft ... 32
Crankshaft assembly - examination ... 11	
Camshaft and camshaft bearings - examination and renovation 12	Engine reassembly - replacing the contact breaker points, advance and retard unit and cover ... 33
Cam followers, cam chain and cam sprockets - examination and renovation ... 13	Engine reassembly - resetting the cam chain tension and tappets ... 34
Cylinder barrel - examination and renovation ... 14	Engine reassembly - setting the points and ignition timing ... 35
Piston and piston rings - examination and renovation ... 15	Replacing the engine in the frame ... 36
Cylinder head - dismantling, examination and renovation ... 16	Starting and running the rebuilt engine ... 37
Dismantling the gearbox - removing and replacing the mainshaft and countershaft bearings and oil seals ... 17	Checking the compression ... 38
Gearbox components - examination and renovation ... 18	Fault diagnosis - engine ... 39
Gearbox components general - examination and renovation ... 19	Fault diagnosis - clutch ... 40
Clutch and primary drive - examination and renovation ... 20	Fault diagnosis - gearbox ... 41
Kickstarter mechanism - examination and renovation ... 21	

Specifications

Engine	CB100, CL100, SL100	CB125S, CD125S, SL125
Bore and stroke	1.988 x 1.949 in (50.5 x 49.5 mm)	2.205 x 1.949 in (56 x 49.5 mm)
Compression ratio	9.5 : 1	9.5 : 1
Displacement	6.04 cu in (99 cc)	7.44 cu in (122 cc)
Horse power	11.5 ps/11,000 rpm	12.0 ps/9000 rpm
Contact breaker points gap	0.012 - 0.016 in (0.3 - 0.4 mm)	0.012 - 0.016 in (0.3 - 0.4 mm)
Spark plug gap	0.024 - 0.028 in (0.6 - 0.7 mm)	0.024 - 0.028 in (0.6 - 0.7 mm)
Valve tappet clearance	0.002 in (0.05 mm)	0.002 in (0.05 mm)

Cylinder bore diameter
  CB100, CL100, SL100
    Standard value ... 50.50 - 50.51 mm (1.9881 - 1.9885 in)
    Serviceable limit ... 50.6 mm (1.992 in) max
  CB125S, CD125S, SL125
    Standard value ... 56.00 - 56.01 mm (2.2047 - 2.2051 in)
    Serviceable limit ... 56.1 mm (2.2086 in) max

Piston diameter
  CB100, CL100, SL100
    Standard value ... 50.47 - 50.49 mm (1.987 - 1.988 in)
    Serviceable limit ... 50.3 mm (1.980 in)
  CB125S, CD125S, SL125
    Standard value ... 55.97 - 55.99 mm (2.2035 - 2.2043 in)
    Serviceable limit ... 55.80 mm (2.1968 in)
Piston ring side clearance
  Standard value ... 0.025 - 0.030 mm (0.0008 - 0.0011 in)
  Serviceable limit ... 0.7 mm (0.0275 in)
Piston ring gap
  Top and second rings
    Standard value ... 0.15 - 0.35 mm (0.0059 - 0.0138 in)
    Serviceable limit ... 0.5 mm (0.0197 in) max
  Oil ring
    Standard value ... 0.15 - 0.04 mm (0.0059 - 0.0158 in)
    Serviceable limit ... 0.5 mm (0.0197 in) max
Valve stem diameter
  Inlet
    Standard value ... 5.450 - 5.465 mm (0.2145 - 0.2150 in)
    Serviceable limit ... 5.420 mm (0.2130 in) min
  Exhaust
    Standard value ... 5.430 - 5.445 mm (0.2138 - 0.2146 in)
    Serviceable limit ... 5.400 mm (0.2126 in) min
Valve spring length
  CB100, CL100, SL100
    Outer
      Standard value ... 40.4 mm (1.591 in)
      Serviceable limit ... 39.0 mm (1.535 in) min
    Inner
      Standard value ... 35.7 mm (1.406 in)
      Serviceable limit ... 34.5 mm (1.358 in) min
  CB125S, CD125S, SL125
    Outer
      Standard value ... 40.9 mm (1.610 in)
      Serviceable limit ... 39.5 mm (1.555 in) min
    Inner
      Standard value ... 35.5 mm (1.318 in)
      Serviceable limit ... 32.0 mm (1.259 in) min

Torque wrench settings	kg m	ft lbs
Engine		
Cylinder head	1.8 - 2.0	11.5 - 14.5
Spark advance	0.8 - 1.2	5.8 - 8.7
Cam sprocket	0.8 - 1.2	5.8 - 8.7
Cylinder mount bolt, 6 mm	1.2 - 1.8	8.7 - 13.0
Left crankcase cover	0.8 - 1.2	5.8 - 8.7
AC rotor	2.6 - 3.2	18.8 - 23.2
AC generator mounting screw	0.8 - 1.2	5.8 - 8.7
Cam chain tensioner arm	0.8 - 1.2	5.8 - 8.7
Right crankcase cover screw	0.8 - 1.2	5.8 - 8.7
Oil filter cover screw	0.8 - 0.4	2.2 - 2.9
Oil filter (locknut, 16 mm)	4.0 - 5.0	29.0 - 36.0
Oil pump gear cover bolt	0.4 - 0.6	2.9 - 4.4
Clutch mounting bolt	0.8 - 1.2	5.6 - 8.7
Gear shift drum stopper bolt	0.8 - 1.2	5.8 - 8.7
Gear shift drum cam bolt	0.8 - 1.2	5.6 - 8.7
Frame		
Front spindle nut	4.0 - 5.0	29.0 - 36.0
Rear spindle nut	4.0 - 5.0	29.0 - 36.0
Rear fork pivot bolt	3.0 - 4.0	21.7 - 29.0
Engine mounting bolt	2.0 - 2.5	14.5 - 18.8
Handlebar mounting bolt	0.9 - 1.1	6.50 - 7.95
Steering stem nut	6.0 - 8.0	43.3 - 57.8
Front cushion mounting bolt	4.0 - 5.0	29.0 - 36.0
Rear cushion mounting nut	3.0 - 4.0	21.7 - 29.0
Torque link mounting bolt	2.0 - 2.5	14.5 - 18.0
Top bridge locknut	4.0 - 4.8	29.0 - 34.7
Final driven sprocket	2.0 - 2.5	14.5 - 18.0
Seat mounting bolt	2.0 - 2.5	14.5 - 18.0

Chapter 1/Engine, clutch and gearbox

Standard parts		
Bolt hex. 6 mm	0.8 - 1.2	5.8 - 8.7
Screw cross, 6 mm	0.8 - 1.2	5.8 - 8.7
Nut, 6 mm	0.8 - 1.2	5.8 - 8.7
Screw cross, 5 mm	0.3 - 0.4	2.2 - 2.9

Engine/gearbox unit - SL 125

Type	Air-cooled, single cylinder, 4 stroke Ohc cylinder inclined 15° from vertical
Bore and stroke	56 x 49.5 mm (2.204 x 1.949 in)
Cylinder capacity	122 cc (7.44 cu in)
Compression ratio	9.5 : 1
Carburettor	Piston valve type. Venturi diameter 22 mm
Ignition	Battery and HT coil
Lubrication system	Wet sump. Pressure fed
Oil capacity	1.0 litre (1.75 imp pints)
Starting method	Kickstart
Spark plug	NGK Type D8 ES
Clutch	Wet. Multi-plate type
Generator	AC type 0.045 kw @ 10,000 rpm
Battery capacity	6V - 6 AH
Air filter	Oiled polyurethane foam
Transmission	5-speed constant mesh
Primary reduction	4.055
Gear ratios:	
1st	2.769
2nd	1.722
3rd	1.272
4th	1.041
5th	0.841
Final reduction	3.266 Drive sprocket 15T Driven sprocket 49T
Horsepower	12 ps @ 9000 rpm

Clutch

Thickness of friction plates	2.9 mm (0.114 in)
Serviceable limit	2.6 mm (0.102 in)

Gearbox

	CB125S	CD125S	SL125	CB100	SL 100
Primary reduction	4.055	4.055	4.055	4.055	4.055
Final reduction	3.267	2.800	3.267	2.857	3.142
Gear ratio:					
1st	2.500	2.769	2.769	2.500	2.500
2nd	1.722	1.722	1.722	1.722	1.722
3rd	1.333	1.272	1.272	1.333	1.333
4th	1.083	1.000	1.000	1.083	1.083
5th	0.923	—	0.815	0.923	0.923

1 General description

The engine is of the all-alloy unit construction type with gear primary drive. A common oil supply is used for the engine and gearbox which enables the engine unit to be kept small and compact. In common with most Japanese machines, an overhead camshaft is employed, driven by a small, but strong, chain which in turn allows the engine to be revved to higher limits than the old conventional pushrod type.

2 Operations with the engine in the frame

Because of the compactness of the engine in the frame it is not possible to remove either the cylinder head or the barrel or to work on the gearbox without first removing the engine from the frame. Work can, however, be carried out on the oil pump and filter, clutch and gearchange mechanism with the engine unit in situ. Refer to the relevant sections of this Chapter.

3 Removing the engine/gearbox unit

1 Before attempting to take the unit out of the frame it is recommended that the machine should be cleaned with Gunk or a similar cleaning agent.
2 Place the machine on the centre stand on firm, level ground. If it is at all unstable use wood or some other firm packing to steady the machine.
3 Using a catchment tank of at least two pints capacity, remove the drain plug from the underside of the motor and allow the oil to drain thoroughly. If it does not drain completely a deluge of oil is released when the side covers are removed later. Allow the oil to drain whilst the rest of the stripdown is undertaken.
4 Turn the petrol tap to 'off' position and pull off the petrol feed pipes.
5 Unfasten the seat fixing bolts at either side of the rear seat, lift up the rear of the seat and pull to the rear to unhook it. Place the bolts back in their holes and screw them in finger-tight to avoid misplacing them.
6 It is always good policy to have various clean boxes to hand when a stripdown is to be carried out, and wherever possible to screw back in place, finger-tight, bolts and nuts after a component has been removed.
7 Remove the battery connections and the battery, then put it in a safe place. If it is being removed for some period of time, a

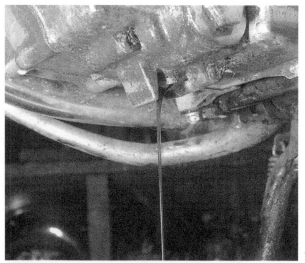
3.3 Remove drain plug and allow oil to drain

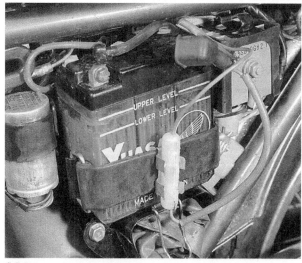
3.7 Remove battery after disconnecting leads

3.8a Remove spring clip to release petrol pipe

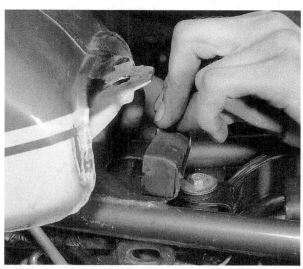
3.8b Ease back rubber block and ....

3.8c .... Lift tank at rear

3.9 Coil is released with head steady

3.10a Removal of the collets under the flange connection

3.10b Remove the swinging arm nut to free the silencer

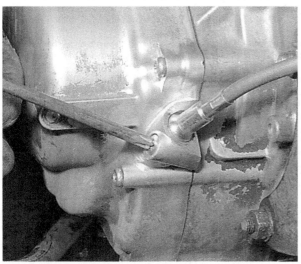
3.11 Release screw to free rev counter cable.

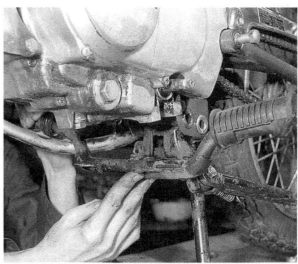
3.12 Footrests bolt to underside of engine unit

3.14 Carburettor is of the flange mounting type

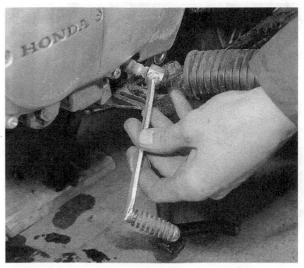
3.16 Remove gear change pedal to aid engine removal

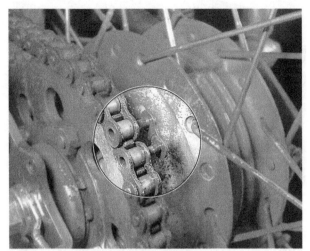

3.18 Remove chain link to separate chain

3.20 Remove the neutral indicator wire by pulling wire from clamp

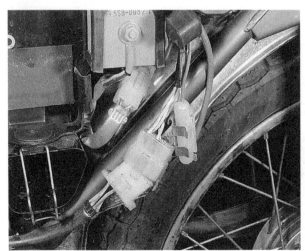

3.21 Unplug engine wiring at block connector

1 amp trickle charge should be given at frequent intervals to keep the battery in good condition.

8 Remove the petrol tank, ease off the spring clip and pull the petrol pipes away from the tap. Ease back the rubber block situated at the rear of the tank and lift the back up whilst pulling it diagonally away from the front mounting blocks. Take care not to lose the blocks, which are plugged into the frame.

9 The next component to remove is the head steady which comprises two triangular plates bolted to the engine and the frame. They also provide an extra point for mounting the engine. Remove the bolts and the coil will also come free. When this is removed, pull off the plug cap and place these items out of the way.

10 The exhaust system is next to be removed. Unfasten the two nuts holding the finned flange to the cylinder head, around the exhaust pipe, and then slacken and remove the largest nut on the end of the swinging arm spindle to free the silencer bracket. Remove the whole system by shaking it whilst maintaining a pull. Two half round collets will fall away from the cylinder head as the exhaust system frees. Put these parts out of harm's way.

11 Undo the crosshead screw which clamps the end of the rev counter cable into the engine (located at the front, right hand top of the engine) and place the cable well out of the way.

12 Remove the bolts from underneath the engine unit which hold the footrest bar in position.

13 Loosen the clutch cable adjuster until enough slack is gained to disconnect it from the engine.

14 Undo the two nuts which hold the carburettor to the cylinder head. Pull off the air cleaner tube and tape it out of the way with the carburettor.

15 Engine removal is helped by taking off the ignition switch and taping it out of the way.

16 Remove the bolt holding the gearchange pedal to the splined shaft. Ease off the pedal and replace the bolt to prevent it being lost.

17 Remove the screws holding the rear left hand crankcase cover, which masks the final drive sprocket.

18 Ease off the spring clip on the joining link and disconnect the rear chain.

19 Unfasten the bolts which hold the chainguard to the machine and remove this part.

20 Using a crosshead screwdriver undo the small screw located in front of the final drive sprocket and pull out the wire from the clamp. This is the neutral indicator light switch.

21 Unplug the block of wires coming from the engine. As they can only go back one way it is unnecessary to mark them.

22 Now undo the six remaining engine bolts and remove all but the last top one at the rear of the engine unit, which should be left in position. At this stage it is wise to have some assistance as the unit is reasonably weighty for its size. With the other person taking the weight of the engine, pull out the last bolt and ease the motor out of the frame.

4 Dismantling the engine - general

1 Before commencing work on the engine unit the external surfaces should be cleaned thoroughly. A motorcycle engine is very prone to grit and any other foreign matter which will find its way into even the smallest of crevices.

2 Clean with Gunk, Jizer or other solvent cleaner, but always make sure that the water cannot get into the engine through the inlet or exhaust ports when washing down.

3 Never use undue force to remove any part. If a component will not come off there is usually a good reason why. Often the wrong dismantling sequence has been used.

4 Note that when the engine is dismantled the gearbox must be dismantled too.

Chapter 1/Engine, clutch and gearbox

3.22a Take the front engine plate off completely

3.22b The best way to ease out the engine

5.1 Remove two cross-head screws to free contact breaker cover

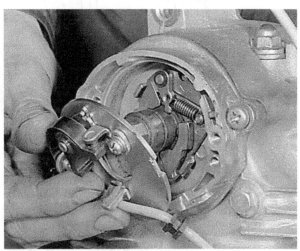
5.3 The advance and retard unit exposed

5.6 Remove the two bolts to free sprocket

5 Dismantling the engine - removing the cylinder head, barrel and piston

1 Remove the two crosshead 5 mm screws which retain the contact breaker points cover.
2 Undo two more crosshead screws which clamp the points baseplate in position.
3 After the baseplate has been removed this will expose the advance and retard mechanism which can be removed by undoing the 6 mm centre bolt and pulling the unit off the end of the camshaft.
4 When the unit has been removed, ease out the location pin for the advance and retard unit if it is loose and put safely to one side; if it is not loose, leave it in position.
5 Next undo the two 6 mm crosshead screws and remove the points casing which also houses a camshaft oil seal. Tap lightly with a rawhide mallet if necessary. The oil seal should be removed and renewed if it has been leaking.
6 Position the engine in top dead centre position and undo the two 6 mm bolts that locate the cam pinion on the camshaft, and remove the cam sprocket from the chain.
7 Using a finger to hold the cam chain up out of the way, extract the camshaft by wriggling it gently then pulling it free.
8 The cam chain can be allowed to drop at this point.

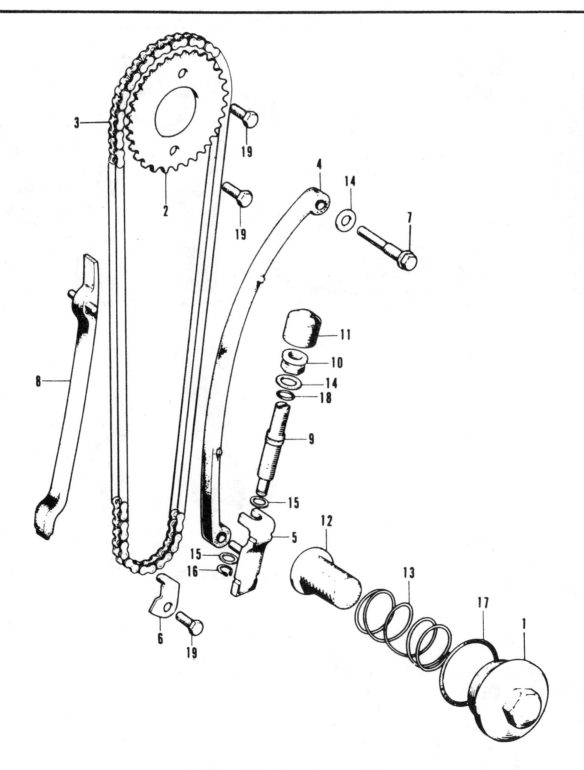

Fig 1.1 Camshaft chain tensioner and oil filter

1 Filter gauze cap
2 Camshaft chain sprocket (32 teeth)
3 Camshaft chain
4 Camshaft chain tensioner
5 Tensioner hinge
6 Camshaft chain guide plate
7 Tensioner pivot bolt
8 Camshaft chain guide
9 Tensioner adjusting bolt
10 Tensioner adjusting nut
11 Tensioner adjusting cap
12 Oil filter gauze
13 Spring
14 Plain washer - 2 off
15 Plain washer - 2 off
16 Circlip
17 O ring
18 O ring
19 Hexagon bolt - 3 off

Chapter 1/Engine, clutch and gearbox

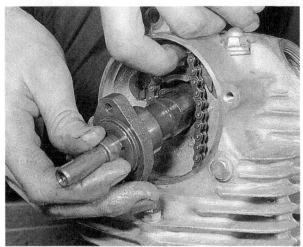

5.7 Ease out the camshaft

5.10 Removing the cam chain tensioner bolt

5.12 Cam chain guide ready for removal

5.14 Piston circlip removal

9 Remove the dome nuts and the 6 mm bolt which holds down the cylinder head.
10 Undo the remaining bolt which secures the top of the cam chain tensioner to the cylinder head.
11 The cylinder head can now be lifted free by gently pulling whilst moving it from side to side.
12 Remove the chain guide from the cylinder barrel, located at the front of the chain slot.
13 The barrel should now lift off but if it is tight, a few light taps with a soft-headed hammer should be sufficient to dislodge it. Just before the barrel releases the piston rings, put a piece of clean rag around the bottom of the piston to stop it becoming damaged. If a piston ring is broken then the rag will catch the bits which may otherwise shower into the crankcase.
14 Support the piston firmly in one hand and with a pair of long-nosed pliers, extract one of the circlips which retain the gudgeon pin in the piston bosses. Press out the pin by pushing it from the opposite side to which the circlip has been removed and lift off the piston when it is clear. Always renew circlips when rebuilding an engine.

6 Dismantling the engine - removing the alternator, cam chain and final drive sprocket

1 With the engine lying on its right hand side, remove the crosshead screws which retain the alternator cover and pull the cover away.
2 Remove the stator plate fixing screws and take it off, not forgetting to slacken the cable clamp.
3 Unfasten the bolt in the centre of the rotor, then using Honda Rotor Puller No 07011 - 03001, screw in and extract the rotor. If an extractor is not available, a large metric threaded bolt or an old metric wheel spindle is just as effective.
4 Undo the bolt which holds the guide for the cam chain and ease out all the components and put them in a clean, safe place - as should be done with all components.
5 The cam chain can now be taken off.
6 The final drive sprocket is easy to remove. Unfasten the two 6 mm bolts and twist the locking plate until it comes away, then lift off the sprocket.

6.2 Slacken cable clamp before lifting off stator plate

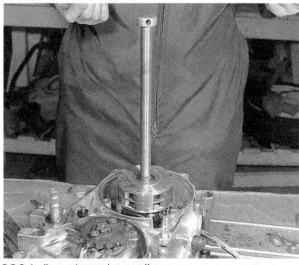
6.3 Spindle can be used as a puller

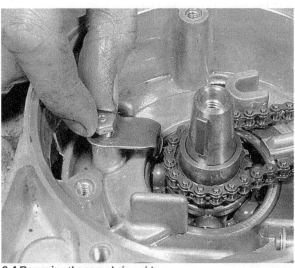
6.4 Removing the cam chain guide

6.6 Final drive sprocket is retained by a plate and two bolts

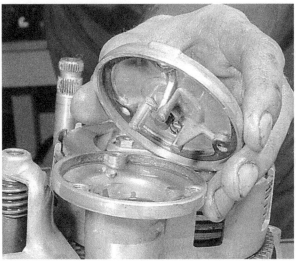
7.2 Lift off the end cover of the centrifugal oil filter

7.3a Fabricated "T" spanner ground or filed from old tube

Chapter 1/Engine, clutch and gearbox

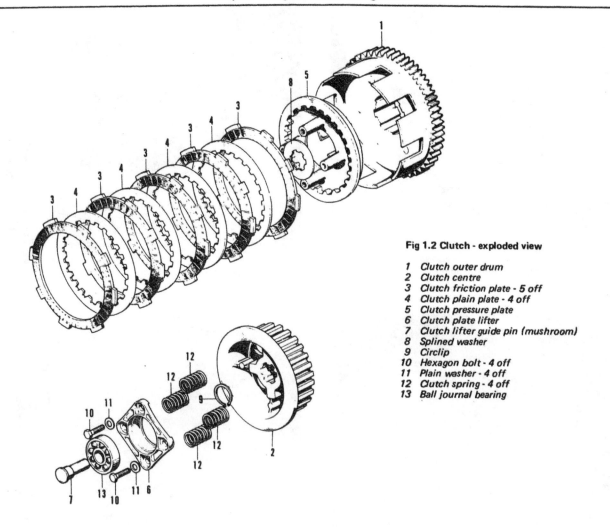

Fig 1.2 Clutch - exploded view

1. Clutch outer drum
2. Clutch centre
3. Clutch friction plate - 5 off
4. Clutch plain plate - 4 off
5. Clutch pressure plate
6. Clutch plate lifter
7. Clutch lifter guide pin (mushroom)
8. Splined washer
9. Circlip
10. Hexagon bolt - 4 off
11. Plain washer - 4 off
12. Clutch spring - 4 off
13. Ball journal bearing

7.3b How the washer is marked

7 Dismantling the engine - removing the centrifugal oil filter, oil pump, clutch and gear selector mechanism

1 Unfasten all the crosshead screws around the periphery of the clutch housing, tap with a soft hammer to break the joint, and lift away.
2 Next remove the three crosshead screws on the small casting on the end of the crankshaft which is the centrifugal oil filter and remove the cover.
3 Using Honda Service Tool No 07086-28301, bend back the tab washer and undo the centre sleeve nut. If the T handle tool is not available, one can easily be made with a piece of stout tube filed to leave two prongs which will locate with the nut, thus enabling the extraction of the centrifugal oil filter case. (Take note that the washer under the nut must be fitted one way - it is marked.)
4 The oil pump can now be disassembled. Unfasten the screws holding on the tachometer drive to the engine, and remove it, then lift out the tachometer drive gear.
5 Lift out the drive gear from the oil pump body.
6 Undo the screws which retain the oil pump body and lift it away. Do not lose the two O rings behind the pump casting.
7 The clutch is removed by undoing the four 6 mm bolts and lifting off the centre. It is important that each bolt must be

7.6 Don't lose 'O' rings at rear of oil pump

7.7 Removing the thrust plate

7.8 Release centre circlip to free clutch plate housing

7.14 Gauze filter should be cleaned and replaced

8.1 A few light taps may be needed to separate crankcases

8.3 Lift out the gear cluster as a complete unit

screwed out one turn at a time so that all four gradually share the load of the spring.

8 Using circlip pliers, ease off the central circlip, thus releasing the outer casting which retains the clutch plates.

9 Pull out the plates (all of them) and lift out the clutch centre with the four pillars attached.

10 Lift out the thrust washer and then the clutch drum can be removed.

11 The gear pinion on the splined crankshaft can now be lifted off.

12 Assuming the gear lever has been taken off, the selector and shaft can now be pulled through.

13 Unfasten the two 6 mm bolts that retain the cam and the stopper and take off both items.

14 The gauze filter should be taken out and cleaned whilst the engine is dismantled. It is located under the large nut (similar to a rocker box cover in appearance) on the left hand side of the engine unit, under the alternator housing.

8 Dismantling the engine - separating the crankcases, removing the crankshaft assembly and gear cluster

1 After removing all the necessary components as described in the preceding Sections, undo the eleven bolts that hold the crankcases together. One is mounted on the right hand crankcase. Tap with a soft headed hammer and the crankcases should come apart. Lift off the left hand crankcase leaving the gearbox components in place with the crankshaft assembly.

2 Gently tap the ends of the crankshaft to displace it from the crankcase.

3 After the crankcases have been separated support the gear cluster in one hand and using a soft headed hammer, tap away the right hand crankcase thus leaving the cluster free.

9 Examination and renovation - general

1 Now that the engine is stripped completely, clean all the component parts in a petrol/paraffin mix and examine them carefully for signs of wear or damage. The following sections will indicate what wear to expect and how to remove and renew the parts concerned, when renewal is necessary.

2 Examine all castings for cracks or other signs of damage. If a crack is found, and it is not possible to obtain a new component, specialist treatment will be necessary to effect a satisfactory repair.

3 Should any studs or internal threads require repair, now is the appropriate time. Care is necessary when withdrawing or replacing studs because the casting may not be too strong at certain points. Beware of overtightening as it is easy to shear a stud by doing this and so giving rise to further problems, especially if the stud bottoms.

4 Where internal threads are stripped or badly worn, it is preferable to use a thread insert, rather than tap oversize. Most dealers can provide a thread reclaiming service by the use of Helicoil thread inserts. They enable the original component to be re-used.

10 Main bearings and oil seals - examination and renovation

1 When the bearings have been washed thoroughly in a petrol/paraffin mix if there is any play in the bearing, or roughness, new replacements must be fitted or if the bearing has at any time been turning on the crankshaft. If the bearing is still fit for use, a proprietary sealant such as Bearing Locktite can be used to take up the slack. To extract the main bearings it is wise to contact your local dealer who will be able to advise you as these bearings are extremely tight fitting.

2 Oil seals should be renewed even though this may seem an unnecessary expense at the time. There is nothing worse than a rebuilt motor leaking oil because an oil seal was not renewed.

11 Crankshaft assembly - examination

1 The crankshaft is a single component which means that if the big end is badly worn or in some cases, has been starved of oil and has broken up, a new assembly will be necessary.

2 Wash well in clean petrol/paraffin mix and if no roughness or slack can be felt re-oil and keep the crankshaft very clean until it is needed for reassembly.

3 If there is any doubt in your mind as to whether the big end has gone then consult your local dealer for his opinion because the assembly is extremely expensive.

12 Camshaft and camshaft bearings - examination and renovation

Wash and lightly oil the camshaft bearing surfaces and insert it into the head. This should be a nice running fit; endfloat should not be taken as wear, but side to side or up and down movement or scoring due to the machine being left to run low on oil will, unhappily, mean that it will be necessary to replace the cam, or head, or both. Once again consult your local dealer in this matter.

13 Cam followers (rockers) cam chain and cam sprockets - examination and renovation

1 The cam chain should be renewed if a large mileage or signs of wear are evident, ie very slack pins or discolouration due to loss of lubrication.

2 The cam sprockets usually give no trouble unless the chain has been left too long before replacement, and in this case they will be hooked over or broken. The large top one which bolts on to the cam is the one which usually goes, as it is much softer than the one affixed to the crankshaft, which is pre-set and should outlast the crankshaft.

3 To extract the cam followers from the cylinder head, take out the bolt above the hole from which the cam has been taken - this will free the retainer and allow the pins to be taken out to permit the cam followers on both rocker arms to be inspected. They should be checked on the surface which runs on the cam for signs of pitting or indentation. Small amounts can be oil stoned lightly to alleviate sharp edges, but if the amount of wear is severe, renewal of the rocker arm is necessary.

14 Cylinder barrel - examination and renovation

The cylinder barrel should be washed off in a petrol/paraffin mix and checked for signs of wear, such as any ridge developing in the top uppermost part or scores or deep scratches resulting from ring breakage in the past. If the bore is badly worn or pitted the barrel can be rebored several times. Oversize pistons are available in sizes of: + 0.25, 0.50, 0.75 and 1.00 mm, or in British sizes of: + 0.010 in, 0.020 in, 0.030 in and 0.040 in.

15 Piston and piston rings - examination and renovation

1 Attention to the piston rings can be overlooked if a rebore is undertaken. It will then be necessary to fit an oversize piston and rings.

2 The piston should be washed off and checked for cracks, score marks, ovality or worn rings or grooves. Usually if any of these are present, to stop the engine compression from by-passing the piston and causing heavy oil consumption, the piston should be renewed and also the rings. Ideally a rebore in one of the previously mentioned sizes should be undertaken.

3 There are various wear limits to refer to as shown in the Specifications Section of this Chapter. The method is simple. Push each ring, after easing it off the piston (be careful not to break the rings as they are extremely brittle) down into the barrel with the piston, until it seats squarely. Use a feeler gauge to measure the end gap, to determine the amount of wear.

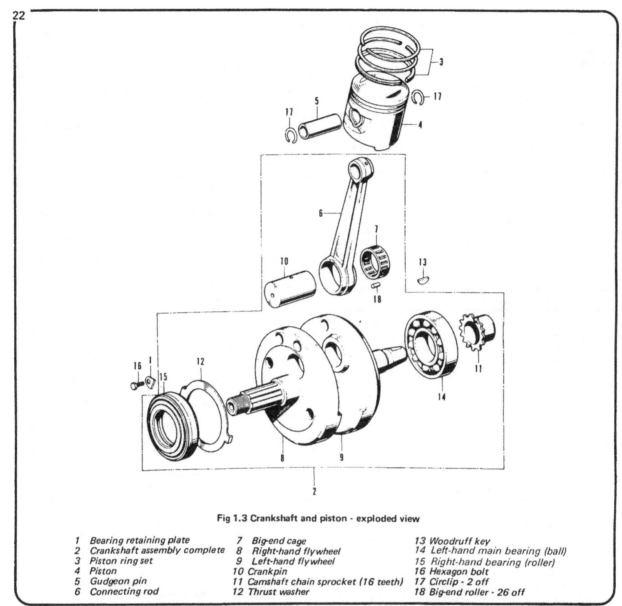

Fig 1.3 Crankshaft and piston - exploded view

1. Bearing retaining plate
2. Crankshaft assembly complete
3. Piston ring set
4. Piston
5. Gudgeon pin
6. Connecting rod
7. Big-end cage
8. Right-hand flywheel
9. Left-hand flywheel
10. Crankpin
11. Camshaft chain sprocket (16 teeth)
12. Thrust washer
13. Woodruff key
14. Left-hand main bearing (ball)
15. Right-hand bearing (roller)
16. Hexagon bolt
17. Circlip - 2 off
18. Big-end roller - 26 off

13.3a The rocker pin retainers must be released ....

13.3b .... to permit rocker removal

Chapter 1/Engine, clutch and gearbox

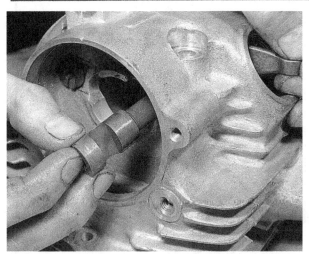

13.3c Pins are relieved by cutaways to miss head and barrel studs

4 Cylinder bore wear can be checked by placing the piston in the cylinder barrel and pushing feeler gauges down the side. This is a somewhat rough and ready check, but if it is carried out properly it can be effective in indicating the extent of bore and piston wear.

5 The piston rings must always be fitted with the markings upwards, so that they can be read from the piston top.

16 Cylinder head - dismantling, examination and renovation

1 The cylinder head itself provides no problems but it should be cleaned off and checked for cracks and distortion. If there is some distortion of the surface where the gasket seat occurs this can be corrected by laying a sheet of very fine wet and dry paper, say 620 grade, on a sheet of plate glass and gently, with a circular motion, rubbing it down until the irregularity has been removed. Do not forget to clean thoroughly after this operation, to remove all the abrasive.

2 It is best to remove all carbon deposits from the combustion chambers, before removing the valves for grinding-in. Use a blunt-ended scraper so that the surface of the combustion chamber is not damaged and finish off with metal polish to achieve a smooth, shiny surface.

3 Before attempting to take out either valve, the cam followers (rockers) will have to be removed.

4 To remove the valves, it is necessary to obtain a valve spring compressor of the correct size, to compress each set of valve springs in turn, so that the split collets can be removed from the valve cap and the valve and valve spring assembly released. Keep each set of parts separate. Note there is a seal around the exhaust valve stem, which should be renewed when the valve is refitted.

5 Before giving the valves and valve seats further attention, check the clearance between each valve stem and the valve guide in which it operates. Some play is essential in view of the high temperatures involved, but if the play appears excessive, the valve guides must be renewed.

6 To remove the old valve guides, heat the cylinder head and drive them out of position with a double diameter drift of the correct size. Refit the new guides, whilst the cylinder head is still warm.

7 Grinding-in will be necessary, irrespective of whether new valve guides have been fitted. This action is needed to remove the indentations in the valve seats caused under normal running conditions by the high temperatures within the combustion chamber. It is also necessary when new valve guides have been fitted, in order to re-align the face of each valve with its seating.

8 Valve grinding is a simple task. Commence by smearing a trace of fine valve grinding compound (carborundum paste) on the valve seat and apply a suction tool to the head of the valve.

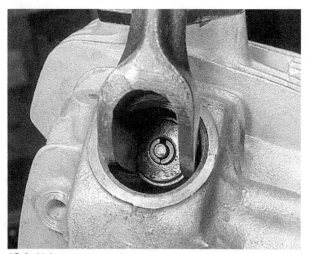

16.4a Valve compressor in use

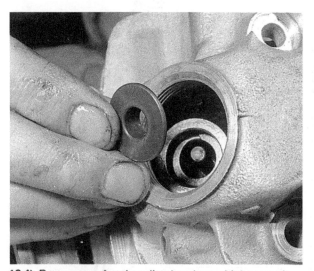

16.4b Remove cap after the collets have been withdrawn and compressor released

16.4c Take out the springs

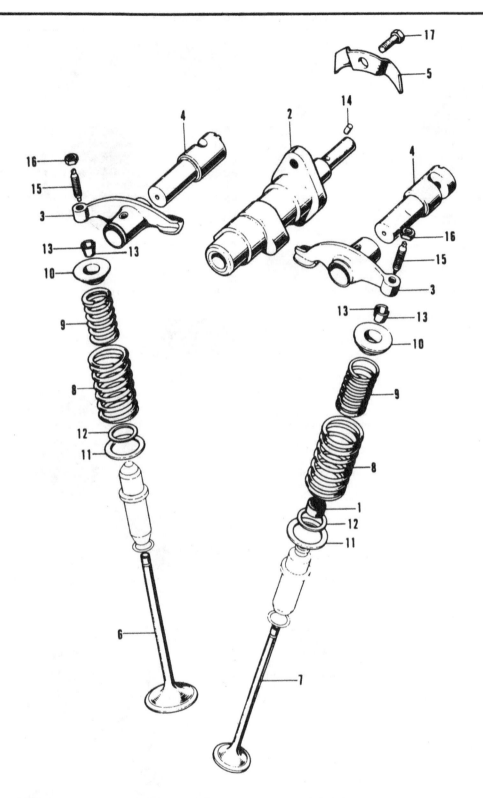

Fig 1.4 Camshaft and valve gear - exploded view

1 Valve stem seal (exhaust valve)
2 Camshaft
3 Rocker arm - 2 off
4 Rocker arm spindle - 2 off
5 Rocker arm spindle retainer
6 Inlet valve
7 Exhaust valve
8 Valve spring - outer - 2 off
9 Valve spring - inner - 2 off
10 Valve spring cap - 2 off
11 Outer valve spring seat - 2 off
12 Inner valve spring seat - 2 off
13 Collets - 4 off
14 Camshaft pin
15 Tappet adjusting screw - 2 off
16 Tappet adjusting nut - 2 off
17 Hexagon bolt

Chapter 1/Engine, clutch and gearbox

Oil the valve stem and insert the valve in the guide so that the two surfaces to be ground in make contact with one another. With a semi-rotary motion, grind in the valve head to the seat, using a backward and forward action. Lift the valve occasionally so that the grinding compound is distributed evenly. Repeat the operation until a ring of light grey matt finish is obtained on both valve and seat. This denotes the grinding operation is complete. Before passing to the next valve, make sure that all traces of compound have been removed from both the valve and its seat and that none has entered the valve guide. If this precaution is not observed, rapid wear will take place due to the abrasive nature of the carborundum base.

9 When deeper pit marks are encountered, or if the fitting of a new valve guide makes it difficult to obtain a satisfactory seating, it will be necessary to use a valve seat cutter and a valve refacing machine. This course of action should be resorted to only in an extreme case, because there is risk of pocketing the valve and reducing performance. If the valve itself is badly pitted, fit a replacement.

10 Before reassembling the cylinder head, make sure that the split collets and the taper with which they locate on each valve are in good condition. If the collets work loose whilst the engine is running, a valve will drop and cause extensive engine damage. Check the free length of the valve springs with the specifications section and renew any that have taken a permanent set.

11 Reassemble by reversing the procedure used for dismantling the valve gear. Do not neglect to oil each valve stem before the valve is replaced in the guide. Do not omit the seal from the exhaust valve stem.

12 Before setting aside the cylinder head for reassembly, make sure that the cooling fins are clean and free from road dirt. Check that no cracks are evident, especially in the vicinity of the holes through which the holding down studs and bolts pass, and near the spark plug threads.

16.8 Grind in both valve seats

18.2a Do not misplace thrust washers

18.2b Examine all splines carefully

17 Dismantling the gearbox - removing and replacing the mainshaft and countershaft bearings and oil seals

1 To see if the bearings need replacing wash them out in a petrol/paraffin mix and check them for roughness, resistance to turning and slack. If any of these defects are evident it is essential that they be replaced, otherwise whining or vibration will occur which in turn will damage other gearbox components.

2 To replace the bearings, heat up the case until it is reasonably hot (this expands the case fractionally, making bearing removal easier). Using a soft drift, ie alloy, brass or copper, tap out the old bearing, and tap in the new whilst the case is still hot, then allow it to cool down.

3 The oil seal, which needs no heating to replace it, should be tapped out and the new one tapped in, taking great care not to damage the soft rubber.

18 Gearbox components - examination and renovation

1 Strip down and reassemble each shaft at a time to avoid any unnecessary confusion. Keep the parts separate and make very careful note of the location of thrust washers and circlips.

2 Examine each of the gear pinions carefully for chipped or broken teeth. Check the internal splines and bushes. Instances have occurred where the bushes have worked loose or where the splines have commenced to bind on their shafts. The two main causes of gearbox troubles are running with a low oil content and condensation, which gives rise to corrosion. The latter is immediately evident when the gearbox is dismantled.

3 The mainshaft and the layshaft should both be examined for fatigue cracks, worn splines or damaged threads. If either of the shafts have shown a tendency to seize, discolouration of the areas involved should be evident. Under these circumstances check the shafts for straightness.

4 Harsh transmission is often caused by rough running ball races especially the mainshaft ball journal bearings. Section 18 of this

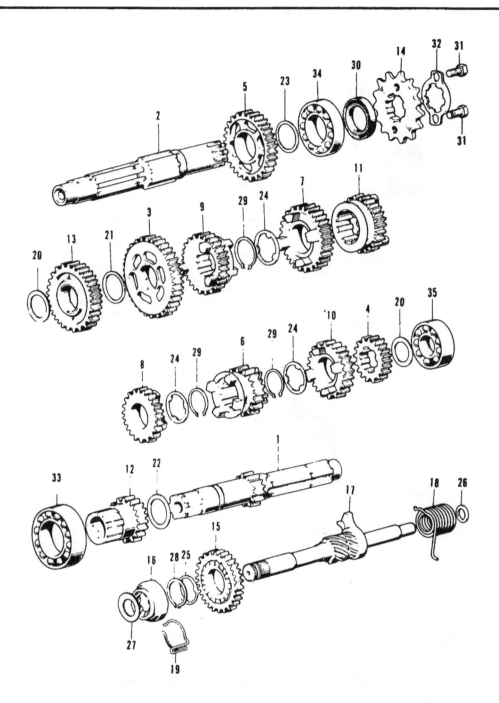

Fig 1.5 Gear cluster and kickstarter - exploded view

1 Gearbox mainshaft
2 Gearbox layshaft (Countershaft)
3 Layshaft 1st gear
4 Mainshaft 2nd gear
5 Layshaft 2nd gear
6 Mainshaft 3rd gear
7 Layshaft 3rd gear
8 Mainshaft 4th gear
9 Layshaft 4th gear
10 Mainshaft 5th gear
11 Layshaft 5th gear
12 Primary kickstarter pinion
13 Primary kickstarter idler pinion
14 Final drive (gearbox) sprocket
15 Kickstarter pinion
16 Kickstarter drive ratchet
17 Kickstarter spindle
18 Kickstarter return spring
19 Starter pinion friction spring
20 Thrust washer - 2 off
21 Thrust washer - 20 mm
22 Thrust washer - 20 mm
23 Thrust washer - 20 mm
24 Thrust washer - 20 mm splined - 3 off
25 Thrust washer - 25 mm
26 Thrust washer - 12 mm
27 Thrust washer - 16 mm
28 Circlip
29 Circlip - 3 off
30 Oil seal
31 Hexagon bolt - 2 off
32 Sprocket retaining plate
33 Ball journal bearing - mainshaft, right-hand
34 Ball journal bearing - layshaft, left-hand
35 Ball journal bearing - mainshaft, left-hand

Chapter 1/Engine, clutch and gearbox

18.2c Make sure no discolouration has occured

18.2d The pinion should be a good sliding fit

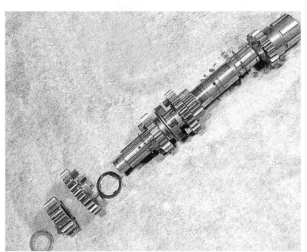

18.2e Shaft assembly laid out for inspection

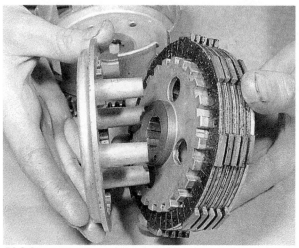

20.3 Check the splines thoroughly

Chapter describes the procedure for removing and replacing the gearbox bearings.

5 All gearbox bearings should be a tight fit in their housings. If a bearing has worked loose and has revolved in the housing, a bearing sealant such as Bearing Loctite can be used, provided the amount of wear is not too great.

6 Check that the selector forks have not worn on the faces which engage with the gear pinions and that the selector fork rod is a good fit in the gearbox housings. Heavy wear of the selector forks is most likely to occur if replacement of the mainshaft bearings is long overdue.

19 Gearbox components general - examination and renovation

1 Before the gearbox is reassembled, it will be necessary to examine each of the components for signs of wear or damage. Each part should be washed in a petrol/paraffin mix to remove all traces of oil and metallic particles which may have accumulated as the result of general wear and tear within the gearbox.

2 Do not omit to check the castings for cracks or other signs of damage. Small cracks can often be repaired by welding, but this form of reclamation requires specialist attention. Where more extensive damage has occurred it will probably be cheaper to purchase a new component or to obtain a serviceable secondhand part from a breaker.

3 If there is any doubt about the condition of a part examined, especially a bearing, it is wise to play safe and renew. A considerable amount of work will be required again if the part concerned fails at a much earlier date than anticipated.

20 Clutch and primary drive - examination and renovation

1 Starting with the largest component first, the clutch outer case should be washed and closely inspected for cracks or loose rivets which hold the drum. If any rivets are loose it is probable that damage has occurred and a new drum will have to be obtained.

2 The next item to check is the clutch pressure plate; this must also be washed and inspected for signs of cracks or distortion. Make sure that the teeth which locate the clutch plates are not badly worn. If they are worn or if the casting is cracked, it must be renewed. If the clutch breaks it will very often completely wreck an engine unit.

3 The clutch centre should also be washed and inspected for cracks, and the inner and outer splines checked for signs of wear or damage. Any small fault can ultimately cause a good deal of

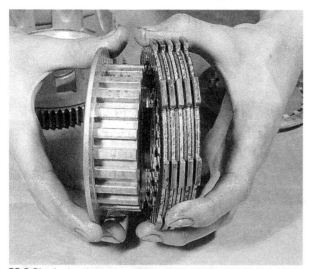
20.6 Check also the mating tongues of the clutch plates

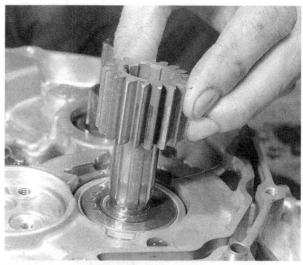
20.9 Examine the gear pinions for chipped or worn teeth

21.1 Check the rotary sliding action of the kickstart assembly

21.2a Examine the assembly carefully for wear or damage

21.2b Don't forget the return spring and renew if weak

22.2a Check the selector drum for worn slots

22.2b Bent or worn selector forks cause gear change problems

22.3 The gear change spindle and arm should not be overlooked

trouble so if there is any doubt it is best to renew at this stage.
4 The clutch lifter plate should also be checked and should be renewed if damaged, worn or cracked.
5 The clutch thrust bearing should be washed out and checked for roughness or slack spots; if it is in order, re-oil and replace it. If roughness or wear is found it should be renewed.
6 The friction plates should be cleaned and checked to see if they are distorted or the bonded lining is coming away. If either of these problems should occur, renewal is essential.
7 The plain plates should be checked for distortion and renewed if necessary.
8 The primary transmission is by gears which should last the life of the motorcycle.
9 In the case of any whining or knocking being heard, check the teeth for chips or burrs; small burrs can be oil stoned away. Cracks in the gear must be looked for if knocking occurs, and if any are found, the gear must be renewed.

21 Kickstart mechanism - examination and renovation

1 Examine closely the condition of the spiral which thrusts the ratchet into gear for signs of pitting or wear, and also check the rotary sliding action. If it jams whilst turned to and fro by hand, it is almost sure to jam under the conditions met when it is in the machine. If it is worn or sticks badly, renewal is called for.
2 If the return spring is broken or weak, replace it and also the circlips. Check closely the ratchet teeth for signs of rounding or chipping and if they are damaged or worn badly, renew them. Both parts of the ratchet will have to be renewed even if only one is worn because the new part will never mate properly with the old.

22 Gearchange selector mechanism - examination and renovation

1 The selector mechanism is simplicity itself, comprising a drum with slots milled in it, the selector forks locating with the slots. When the drum is rotated the selectors slide up and down along the shaft which holds the forks, because the slots are shaped like waves. This places the gears in different positions, each time the drum is rotated.
2 Check for wear in the slots of the drum or the pins that engage with these slots. Do not lose the springs from the selector fork pins if dismantled. A problem which is likely to occur if the gearbox has suffered very hard use is that of bent selector forks, which position the gear incorrectly thus causing the gearbox to malfunction. The only way to treat a worn selector fork or one that is damaged is to renew it.

3 Check also the selector gearchange arm and spindle. If either is bent it will make gear selection difficult and the defective part should therefore be renewed.

23 Engine reassembly - general

1 Before the engine is reassembled, all the various components must be cleaned thoroughly so that all traces of old oil, sludge dirt and gaskets etc are removed. Wipe each part with clean, dry lint-free rag to make sure there is nothing to block the internal oilways of the engine during reassembly.
2 Make sure that all traces of the old gaskets have been removed and that the mating surfaces are clean and undamaged. One of the best ways to remove old gasket cement is to apply a rag soaked in methylated spirit. This acts as a solvent and will ensure the cement is removed without resort to scraping and subsequent risk of damage. Special care should be taken with regard to the crankcases to ensure that the locating dowels are positioned correctly.
3 Gather together all the necessary tools and have available an oil can filled with clean engine oil. Make sure the new gaskets and oil seals are to hand; nothing is more unfuriating than having to stop in the middle or a reassembly sequence because a vital gasket or replacement has been overlooked.
4 Make sure that the reassembly area is clean and that there is adequate working space. Refer to the torque and clearance settings whenever they are given. Many of the smaller bolts are easily sheared if they are overtightened. Always use the correct size screwdriver bit for the crosshead screws and never an ordinary screwdriver or punch.

24 Engine reassembly - replacing the crankshaft assembly

1 Make sure that the main bearings, if they have been disturbed, are up against their shoulders on the crankshaft.
2 Make sure that the unit is scrupulously clean, with no dirt or grit. Clean oil should be applied sparingly on the bearings just before reassembly.

25 Engine reassembly - replacing the gear cluster and selectors

1 Reassembly is aided by the inclusion of several diagrams showing the component parts of the gear cluster.
2 If it has been necessary to take the selector forks from their shaft, replace the selector pins, hold the ball and spring with a screwdriver and slide through the selector shaft.

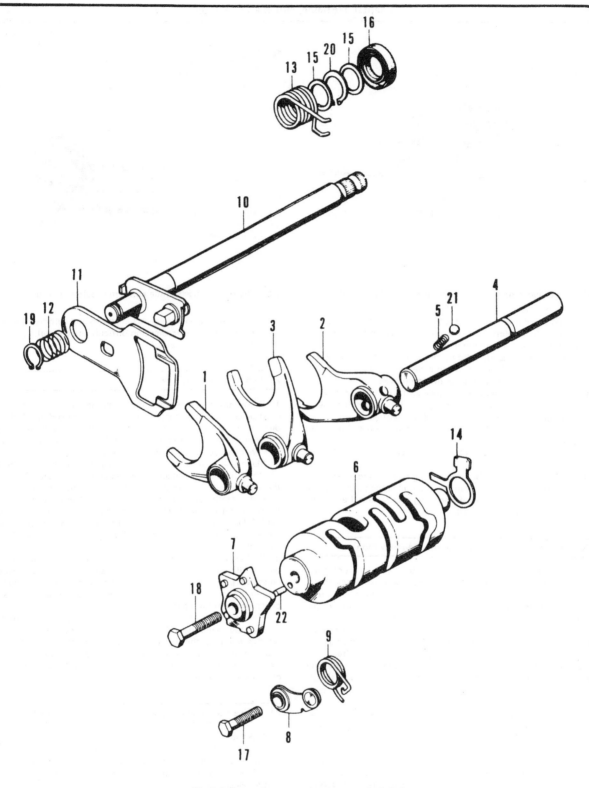

Fig 1.6 Gear selector mechanism - exploded view

1 Right-hand gear selector fork
2 Left-hand gear selector fork
3 Centre gear selector fork
4 Selector fork spindle
5 Spring
6 Gear change drum
7 Gear change cam
8 Gear change stop arm
9 Stop arm spring
10 Gearchange lever spindle
11 Gear change plate
12 Gear change plate spring
13 Gear change return spring
14 Neutral contact rotor
15 Thrust washer - 2 off
16 Oil seal
17 Hexagon bolt
18 Hexagon bolt
19 External circlip
20 External circlip
21 Ball bearing
22 Pin

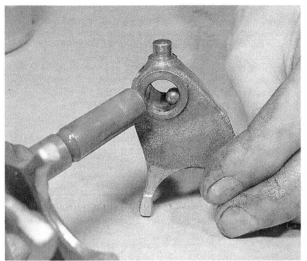
25.2 Hold spring-loaded ball with a small screwdriver when replacing shaft

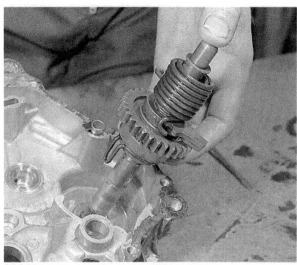
25.3 Make sure end of return spring locates correctly

25.4a Do not omit any spacers or thrust washers when reassembling

25.4b How the selectors and gear trains are assembled

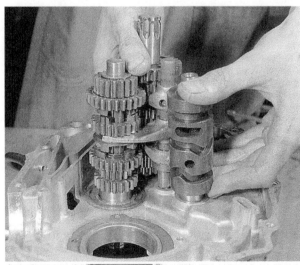
25.6a Fit gear cluster with drum as a complete unit

25.6b Selector forks must engage correctly with their slots

25.6c Don't forget neutral indicator switch tab

26.4 Reassembling the crankcases

27.1a Locate with pin on selector drum and tighten

3 Install the kickstart shaft into the right hand crankcase and insert one end of the kickstart return spring into the hole in the case and the other end around the boss in the case. Check that the spring is tensioned.
4 Reassemble the gears on their shafts, not forgetting the various spacers and shims. Refer to the diagrams if any difficulty occurs when reassembling them on their respective shafts.
5 Install the shafts with the gears and locate them in the right hand crankcase.
6 Locate the selectors and then push the spindle into its hole and fit in the selector drum. Do not forget the neutral switch indicator tab located on the end of the drum.

26 Engine reassembly - joining the crankcases

1 It is important that the gear cluster and selector drum assembly is correctly replaced and that it is both scrupulously clean and well oiled.
2 Fit the crankshaft assembly into the right hand crankcase; no force should be used as this may well cause damage to the main bearings.
3 Very slowly and carefully fit the new gasket. (No compound should be used unless the machined faces are badly marked.)
4 Pick up the left crankcase with both hands (have one last look to make sure that all the components are in their correct places) and lower it firmly and squarely on to the other half. It should just fall into place but may need a slight tap with a soft hammer to make the joint correctly as the dowels locate. Make sure that everything is free to rotate, then replace and tighten all the bolts holding the crankcases together. Check again that all the shafts are free to rotate.

27 Engine reassembly - replacing the clutch, gearchange mechanism, primary drive and oil pump

1 Relocate the pin of the selector cam and tighten the bolt in the centre. Bolt on the cam roller, taking care to hook the roller over the cam before the final tightening.
2 Oil lightly and slide the gearchange shaft through the crankcase, taking care not to damage the oil seals, then line up the selector square with the protruding pins on the cam.
3 Place the primary gear on the end of the crankshaft, line up the splines and push it home.
4 Place two new O rings in the rear of the oil pump, replace the oil pump complete and tighten.
5 Next place the clutch drum on the splined shaft of the gearbox and position the splined washer, making sure that it is properly located.
6 Build up the clutch centre with friction and plain plates, using both hands to align the various splines.
7 Using circlip pliers and a new circlip, position the assembly over the splined clutch centre and release the circlip. Check that it has seated correctly in its groove.
8 Replace the clutch lifting pin in the bearing. Place the bearing in the outer thrust casting.
9 Place the springs over the pillars of the clutch centre; push the four bolts through the casting and line them up, then tighten them gradually until all four are quite tight. Torque to 0.8 - 2.1 kg m (5.8 - 8.7 ft lbs).

28 Engine reassembly - replacing the centrifugal oil filter

1 Having already positioned and located the primary gear, push the outer casting of the filter up tight against the gear pinion.
2 Replace the special washer, noting it is marked to show the way in which it has to be fitted. Follow with the sleeve nut. Tighten thoroughly to a torque setting of 4.0 - 5.0 kg m (29 - 36 ft lbs).
3 Fit a new gasket, then replace the cover and firmly tighten the three crosshead fixing screws.

27.1b Hook roller over cam before tightening

27.2 Selector quadrant in place

27.4 Fit new "O" rings under rear of oil pump

27.6 Assemble clutch centre and lower on to the splines

27.7 Replace the centre circlip

27.8 Assemble bearing in thrust plate

27.9 Place springs over pillars in clutch centre

28.1 Push casting against drive pinion

28.2 Tighten sleeve nut

28.3 Use new gasket for joint

30.2 Replace lower chain tensioner after guide is in position

30.6 "O" ring provides seal for alternator

Chapter 1/Engine, clutch and gearbox

29 Engine reassembly - replacing the right hand engine cover

1 Have one last check to see and make sure that everything is correctly fitted and tightened and that the surfaces are clean.
2 Holding the case firmly in between both hands, having put the new gasket in the correct place, lower the case slowly over the kickstart shaft, taking great care not to damage the seal. Locate the cover on the dowels and if necessary tap it into position with a soft headed hammer or mallet.
3 If the case will not go right on do not force it but pull it off and discover what is causing difficulty. Often the kickstart shaft is pulled slightly out of alignment by the return spring.
4 Place all the screws around the outside of the case and tighten them.

30 Engine reassembly - replacing the cam chain, tensioner and alternator

1 Place the cam chain around the lower sprocket on the left hand end of the crankshaft and lift it up through the slot through which it runs.
2 Replace the chain guide, the rear of the rubber coated chain tensioner and the springs.
3 Replace the Woodruff key in the slot in the end of the crankshaft, to locate the generator rotor.
4 Place the rotor over the shaft, line up the keyway with the key and push it into position. Fit the centre bolt and tighten; torque to a setting of 2.6 - 3.2 kg m (19 - 23.0 ft lbs).
5 Place the stator plate in the correct position around the rotor. Locate the cable grommet which protects the alternator wires and tighten the retaining screws.
6 Place the large rubber O ring around the unit. Replace the cover and tighten.
7 Replace the oil filter gauze and spring in the crankcase and using a new sealing O ring, tighten the cap-like cover firmly.
8 Replace the final drive gearbox sprocket and if it was taken out, the neutral indicator light switch.

31 Engine reassembly - replacing the piston, cylinder barrel and cam chain guide

1 Put some clean rag around the connecting rod and with the largest of the valve cutaways in the piston facing backwards, (the largest cutaway is to clear the inlet valve) and also has 'IN' stamped on it, lightly oil and position the piston over the connecting rod, push through the gudgeon pin and fit the new circlips. Double-check that the circlips are correctly seated.
2 Make sure that the base where the cylinder barrel is to seat is clean, and that the two dowels are in place. Fit a new base gasket and oil the barrel liberally.
3 Support the piston and also the cam chain, feeding the latter up the tunnel as the barrel is lowered in to position, taking care not to force the barrel too hard. By rocking gently, the piston and rings should gradually ease into the barrel.
4 Hold the cam chain in one hand and with the other guide the cam chain guide into place until it is located correctly. Do not drop the cam chain down the tunnel, or you will have to start all over again.

32 Engine reassembly - replacing the cylinder head and camshaft

1 Having placed the three dowels, the O ring and the cylinder head gasket in place, support the cam chain and lower the cylinder head into place.
2 As the cylinder head rests on the cylinder barrel, line up and replace the rear camshaft chain tensioner bolt, taking care to locate the hole in the top of the tensioner with the bolt before tightening.
3 Place the long bolt through the cylinder head, finger-tight.

30.8a Replace final drive sprocket and tighten locking plate

30.8b Don't forget the neutral switch

31.1 Oil gudgeon pin and double check circlips after fitting

31.3 Gently lower the barrel into place after rings have engaged

31.4 Insert the cam chain guide

32.2 Re-insert and tighten the rear tensioner bolt

32.6a "T" mark align correctly

32.6b Dot on the camshft sprocket must align with notch

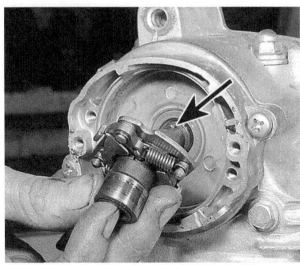
33.3 Locate the advance and retard unit correctly with peg

Chapter 1/Engine, clutch and gearbox

34.2 The cam chain adjuster

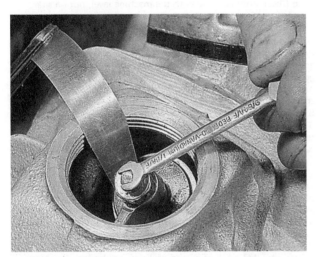

34.6 Method of adjusting the tappets

4 Liberally oil the camshaft and opening the chain out with two finger, slide the camshaft through until it is correctly located in the cylinder head. Position the camshaft with the lobes pointing downwards so that the valves are closed.
5 Rotate the crankshaft slowly until the T mark on the alternator rotor aligns exactly with the index mark on the casing edge. Place the camshaft sprocket over the camshaft and mesh the chain with the sprocket teeth, positioning the sprocket so that the small dot mark adjacent to one tooth is in line with the notch on the top of the cylinder head casting. Locate the two sprocket bolts, turning the camshaft slightly if required, and tighten them fully.
6 Valve timing is now completed, but check the accuracy of the alignment before continuing.
7 Replace the washers and dome nut and tighten all the cylinder head bolts. Torque to 1.8 - 2.0 kg m (11.5 - 14.5 ft lb).

33 Engine reassembly - replacing the contact breaker points, advance and retard unit and cover

1 Replace the O ring and taking care not to damage the oil seal, ease the contact breaker casting into place and tighten.
2 Locate the advance and retard unit on its locating peg, replace in the centre bolt and tighten.
3 Replace the contact breaker baseplate over the camshaft then locate and tighten. Do not fully tighten as the backplate may require moving to alter the timing.

34 Engine reassembly - resetting the cam chain tension and tappets

1 If the cam chain is slack the valve timing will be incorrect and cause poor engine performance. Listen to the chain noise when the engine is ticking over and if a chattering noise is heard the chain is too slack. If the chain produces a whining noise, it is too tight.
2 Loosen the cam chain adjuster locknut and turn the chain adjuster.
3 Adjust the chain tension so that the chain is operating quietly. Do not overtighten as this will cause the whining noise. Turning the adjuster anticlockwise tightens the chain.
4 Do not forget to tighten the locknut after the adjustment has been completed.
5 To set the tappets, take off the tappet covers and the rotor cover and align the T mark with the pointer. Make sure it has lined up on the correct stroke by watching the valves; if they are rocking when the T mark is lined up the engine is in the wrong position. Turn it over again until the T mark lines up and the valves do not move when the rotor is moved slightly. The correct tappet settings are: 0.05 mm (0.002 inch) inlet and exhaust;
6 Having positioned the rotor correctly, slacken the locknut on the rocker arm and insert a 0.05 mm (0.002 inch) feeler gauge between the adjuster and the valve top. With the fingers or a small spanner screw down the square-ended adjuster until the feeler gauge is just able to slide freely, and then, without altering the adjustment, tighten the adjuster locknut. Recheck after tightening. If it has altered, go through the adjustment routine again until it is correct and then repeat with the other valve.
7 Replace the tappet covers and the rotor cover and tighten all three.

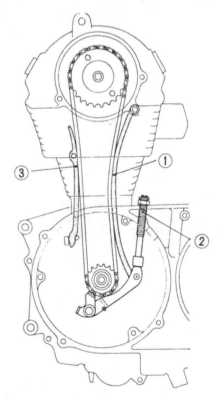

Fig 1.7 Camshaft chain tensioner - mode of operation

1 Cam chain tensioner
2 Adjusting bolt
3 Cam chain guide

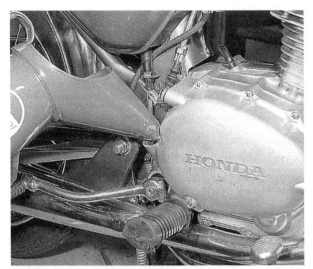

36.9 Do not forget to refill with oil

37.5 Check after running with the machine level, not on side stand

35 Engine reassembly - setting the points and ignition timing

Refer to Chapter 3 for full details of the procedure

36 Replacing the engine in the frame

1 Position the frame so that it stands firmly and place the engine unit in the correct position. Insert the rear engine bolt to hold it in place; do not forget the earth strap is attached to this bolt.
2 Replace the front engine plate and insert all the bolts, but do not tighten any of them yet.
3 Replace the head steady and all the bolts.
4 Tighten all the engine bolts and those of the head steady, then double-check them.
5 Reconnect the electrical block from the alternator and reconnect the wire to the end of the neutral indicator switch and also the wire from the contact breaker points. Replace the battery and connect the leads.
6 Replace and tighten the exhaust system, carburettor, kickstart, gearchange lever and the rev counter cable.
7 Replace the rear chain making sure the spring link is refitted the correct way. Replace the chainguard and tighten the retaining bolts.
8 Replace the fuel tank and attach the petrol pipe, also any extras which may have been fitted to the machine. Replace the seat and reconnect and adjust the clutch cable.
9 Refill with clean, new oil. Quantity: 1 litre (1.05 qt).
10 Have a final check to ensure all the drain plugs and other equipment is tight.

37 Starting and running the rebuilt engine

1 Switch on the ignition and run the engine slowly for the first few minutes, especially if the engine has been rebored.
2 Check that all controls function correctly. Check for any oil leaks or blowing gaskets.
3 Before taking the machine on the road, check that all the legal requirements are fulfilled and that items such as the horn, speedometer and lighting equipment are in full working order. Remember that if a number of new parts have been fitted, some running-in will be necessary. If the overhaul has included a rebore, the running-in period must be extended to at least 500 miles, making maximum use of the gearbox so that the engine runs on a light load. Speeds can be worked up gradually until full performance is obtainable by the time the running-in period is completed.
4 Do not tamper with the exhaust system under the mistaken belief that removal of baffles or replacement with a quite different type of silencer will give a significant gain in performance. Although a changed exhaust note may give the illusion of greater speed, in a great many cases quite the reverse occurs in practice. It is therefore best to adhere to the manufacturer's specification.
5 Check the oil level after the initial run and top up if necessary.
6 To check the clutch adjustment first measure the free play at the end of the handlebar lever. If it exceeds 10–20 mm (2/5 – ¾ in) the clutch must be adjusted in accordance with the following procedure. Loosen the knurled locking nut at the lever and screw the adjuster right in to the body of the lever. Do the same with the cable adjuster on the right-hand engine casing to allow plenty of slack for the main adjustment to the operating mechanism. Loosen the locking nut at the rear lower portion of the right-hand casing and turn the slotted adjusting screw in the centre anti-clockwise until a slight resistance is felt. So that the clutch will not be permanently engaged back off the adjuster by 1/8 to ¼ of a turn and tighten the locknut. Now that the majority of the adjustment has been taken up the cable adjuster on the engine casing should be adjusted. Screw out the bolt until the play at the handlebar lever is within the specified limit. A finer adjustment can be made by screwing the knurled adjuster at the handlebar lever outwards. Remember to tighten the locknut in both cases. Check for proper adjustment by starting the engine, operating the clutch and the gearchange.

38 Checking the compression

1 After the engine has had time to settle down, it is sometimes advisable to check the compression of the engine, especially if performance is not up to standard. The check should always be carried out when the engine has warmed up and only when the various components of a rebuilt engine have had the opportunity to bed down into their correct relationship.
2 A compression gauge is needed for the test. Remove the spark plug and insert the rubber tip of the compression gauge into the spark plug hole so that it forms an effective seal. Open the twist grip wide and operate the kickstarter.
3 If the engine is in good condition, a pressure of 170 psi (12 kg cm^2) should be recorded on the dial of the gauge. A much lower reading indicates leaking valves, defective or sticking piston rings, a blowing cylinder head gasket or incorrect tappet adjustment, causing a valve to stick open. A higher reading is caused by an excessive build-up of carbon in the combustion chamber and on the piston, indicating a decoke is necessary.

Chapter 1/Engine, clutch and gearbox

39 Fault diagnosis - engine

Symptom	Cause	Remedy
Engine starting difficulty or does not start	Lack of compression	
	(1) Valve stuck open	Adjust tappet clearance.
	(2) Worn valve guides	Renew.
	(3) Valve timing incorrect	Check and adjust.
	(4) Worn piston rings	Renew.
	(5) Worn cylinder	Rebore.
	No spark at plugs	
	(1) Fouled or wet spark plug	Clean.
	(2) Fouled breaker contact points	Clean.
	(3) Incorrect points gap	Adjust.
	(4) Incorrect ignition timing	Check and adjust.
	(5) Open or short circuit in ignition	Check and renew fuse.
	No fuel flowing to carburettor	
	(1) Blocked fuel tank cap vent hole	Clean.
	(2) Blocked fuel tap	Clean.
	(3) Faulty carburettor float valve	Renew.
	(4) Blocked fuel pipe	Clean.
Engine stalls whilst running	Fouled spark plug or contact breaker points	Clean.
	Ignition timing incorrect	Adjust.
	Blocked fuel line or carburettor jets	Clean.
Noisy engine	Tappet noise:	
	Excessive tappet clearance	Check and reset.
	Weakened or broken valve spring	Renew springs.
	Knocking noise from piston:	
	Worn piston and cylinder noise	Rebore cylinder and fit oversize piston.
	Carbon in combustion chamber	Decoke engine.
	Worn gudgeon pin or connecting rod small end	Renew.
	Cam chain noise:	
	Stretched cam chain	Renew chain.
	Worn cam sprocket or timing sprocket	Renew sprockets.
Engine noise	Excessive run-out of crankshaft	Renew.
	Worn crankshaft bearings	Renew.
	Worn connecting rod	Renew flywheel assembly.
	Worn transmission splines	Renew.
	Worn or binding transmission gear teeth	Renew gear pinions.
Smoking exhaust	Too much engine oil	Check oil level and adjust as necessary.
	Worn cylinder and piston rings	Rebore and fit oversize piston and rings.
	Worn valve guide	Renew.
	Damaged cylinder	Renew cylinder barrel and piston.
Insufficient power	Valve stuck open or incorrect tappet adjustment	Re-adjust.
	Weak valve springs	Renew.
	Valve timing incorrect	Check and reset.
	Worn cylinder and piston rings	Rebore and fit oversize piston and rings.
	Poor valve seating	Grind in valves.
	Ignition timing incorrect	Check and adjust.
	Defective plug cap	Fit replacement.
	Dirty contact breaker points	Clean or renew.
Overheating	Accumulation of carbon on cylinder head	Decoke engine.
	Insufficient oil	Refill to specified level.
	Faulty oil pump and/or blocked oil passage	Strip and clean.
	Ignition timing too advanced	Re-adjust.

40 Fault diagnosis - clutch

Symptom	Cause	Remedy
Clutch slips	Incorrect adjustment	Re-adjust.
	Weak clutch springs	Renew set of four.
	Worn or distorted pressure plate	Renew.
	Distorted clutch plates	Renew.
	Worn friction plates	Renew.
Knocking noise from clutch	Loose clutch centre	Renew clutch.
Clutch does not fully disengage	Incorrect adjustment	Re-adjust.
	Uneven clutch spring tension	Re-adjust.
	Distorted clutch plates	Renew.

41 Fault diagnosis - gearbox

Symptom	Cause	Remedy
Difficulty in engaging gears	Broken centre gear selector pawl or cam	Renew.
	Deformed gear selector	Repair or renew.
Machine jumps out of gear	Worn sliding gears on main shaft and countershaft	Renew.
	Distorted or worn gear selector fork	Repair or renew.
	Weak gearchange drum stop spring	Renew spring.
Gearchange lever fails to return to normal position	Broken or displaced gearchange return spring	Renew or repair.

Chapter 2 Fuel system and lubrication

Contents

General description ... 1	Throttle cable - examination and replacement ... 8
Petrol tank - removal and replacement ... 2	Air cleaner - removal and replacement ... 9
Petrol tap - removal and replacement ... 3	Exhaust system - general ... 10
Petrol feed pipe ... 4	Engine lubrication - removal and replacement of the oil pump 11
Carburettor - removal ... 5	Locating and cleaning the oil filters ... 12
Carburettor - dismantling, examination and reassembly ... 6	Fault diagnosis - fuel system and lubrication ... 13
Carburettor adjustment - checking the settings ... 7	

Specifications

Carburettor settings

	CB100, CL100, SL100	CB125S, CD125S, SL125S
Main jet	110	105
Air jet	100	100
Needle jet	2.6 x 3.8 length 10	2.6 x 3.8 length 10
Needle jet holder	5.0	5.0
Jet needle	2° 30'' x 3 step 2.495	2° 30'' x 5 step 2.495
Air screw	1 1/2 ± 1/8	1 1/2 ± 1/8
Throttle valve	2.5 cutaway width 1.2 depth 0.2	2.5 cutaway width 1.8 depth 0.2

Slow jet	38 1 0.8 x 2
	     2 0.8 x 2
	     3 0.8 x 2
		38 0.9 x 2 x 4

Fuel level	24 mm (0.9449 in)	24 mm (0.9449 in)

1 General description

The fuel system comprises a petrol tank from which petrol is fed by gravity to the float chamber. It is controlled by a petrol tap with a built-in filter or sediment bowl. The tap has three positions: 'Off', 'On', 'Reserve', the latter providing a reserve supply of petrol when the main supply has run out. For cold starting the carburettor has a choke (manually operated) which is operated at the rider's discretion. The machine should run on 'Choke' for the least amount of time.

The lubrication system is of the pressure fed type, supplying oil to almost every part of the engine. There is a centrifugal filter mounted directly on the end of the crankshaft. Centrifugal force caused by the rotation of the engine throws the heavier impurities outwards where they stick to the walls, allowing only the clean, lighter oil through. Oil is picked up by the oil pump and pressure fed through the right hand crankcase where it is diverted into two routes. In one direction it goes through a passage in the right hand crankcase cover and then through the oil filter to the crankshaft. The other direction takes the oil through a passage via a cylinder head stud to the camshaft and rocker arms. The transmission also receives oil under pressure, relying upon this simple yet very efficient system.

2 Petrol tank - removal and replacement

1 The fuel tank is not bolted to the machine in any way. It is held in place by three rubbers; two at the inner front and one at the rear under the seat, which it is necessary to remove.
2 Unfasten the bolt on each side of the rear of the seat. Lift up the back a little and pull back until the seat disengages with its location bracket and lifts clear.
3 Turn the fuel to 'Off' position and ease off the rubber fuel feed pipe clip and pipe. Lift the rear of the tank, pull to the rear, then lift away.
4 If difficulty is found in replacing the tank, apply a small amount of petrol to the tank front rubbers before reassembly. Follow the removal procedure in reverse order.

3 Petrol tap - removal and replacement

1 Remove the petrol tank, drain and unscrew the sediment bowl that threads on to the petrol tap.
2 Using a crosshead screwdriver, remove the tap from the tank and replace the seal if necessary. Replace in reverse order.

5.2a Remove the right hand side cover

5.2b Disconnect the rubber air hose from the carburettor intake

5.3 Slide assembly exposed

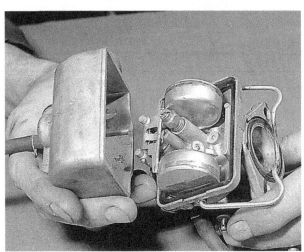
6.2a Float assembly exposed

4 Petrol feed pipe

As soon as the pipe becomes perished or shows signs of cracking or going hard it must be replaced with a new one cut to the same length as the original.

5 Carburettor - removal

1 Pull off the clip and the petrol feed pipe from the carburettor after turning off the fuel.
2 Remove the right hand side cover. Undo the air cleaner hose clip from the carburettor intake.
3 Unscrew the top of the carburettor and pull out the slide on the end of the throttle cable.
4 Undo the two 6 mm nuts holding the carburettor to the cylinder head and remove the carburettor complete with overflow tube.

6 Carburettor - dismantling, examination and reassembling

1 Before stripping the carburettor it is advisable to lay clean paper on the workbench as there are small components in the carburettor that may be lost.
2 Having removed the carburettor from the machine, flick back the spring clip and take off the float bowl. The correct float level is needed for the carburettor to function correctly, and this must be set or checked with a gauge. Contact your local Honda dealer in this instance. The float height is 24 mm (0.826 inch).
3 Unscrew the main jet with a screwdriver and check that it is clean. Blow through all airways to remove any particles of dirt or fluff. Always clean with air and never use a pointed instrument, which may enlarge the jet.
4 To adjust the needle height (this should only be changed with a qualified mechanic's advice) hold the slide in one hand and pull the spring out of the slide insert. The needle can then be taken out. The needle has several slots in it; to make the mixture weaker the needle should be lowered and to make it richer it should be raised.
5 After cleaning and removing any foreign bodies from the carburettor, rebuild in reverse order.

Chapter 2/Fuel system and lubrication

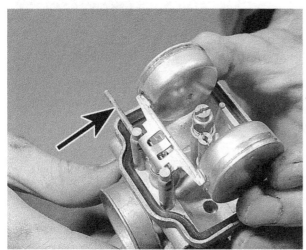
6.2b To remove float assembly extract pivot pin

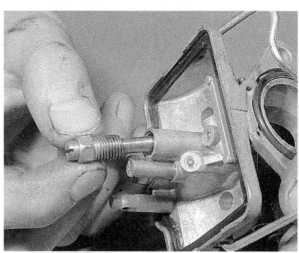
6.3 To clean jet holder unscrew large nut

6.4 Adjust needle height by moving clip in needle notches

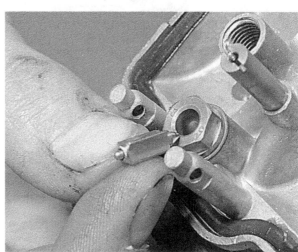
6.5a Prior to reassembly check float needle and seat for wear

6.5b Do not forget to replace plug lead clip when retitting carburettor

7 Carburettor adjustment - checking the settings

1 It is necessary before checking the carburettor settings that the engine is warmed up to its normal running temperature.

2 After this set the mixture. Screw in the brass screw in the centre of the side of the carburettor (adjacent to the slide) to make the tickover faster - about 1000 - 1500 rpm. Then using a screwdriver in the other brass screw (mixture screw) screw it in and out very slowly and leave it where the engine revs are highest. To weaken the mixture, unscrew. To richen the mixture - screw inwards.

3 Check the sizes of the various jets etc with the Specifications Section of this Chapter. These valves have been established by the manufacturer after expensive tests and it will not be possible to improve on them.

Chapter 2/Fuel system and lubrication

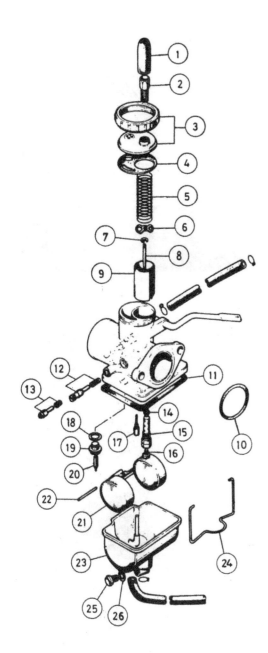

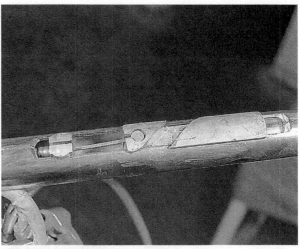

8.3 Bendix throttle system in location

8 Throttle cable - examination and replacement

1 The throttle cable should not be overlooked because it is in constant use and if it should break in a remote spot the machine is immobilised. To keep it in good working order oil the throttle cable regularly and make sure that it has a good 'run' ie no tight bends.

Replacement of the SL series cable

2 There are two different types of twist grip. The one fitted to the SL125 trail series is a conventional drum type and to change the cable it is necessary to unscrew the carburettor top and detach the cable from the slide, then undo the screws from underneath that hold the twist grip control together. The twist grip drum can then be eased away and the cable nipple detached. Finally, unscrew the cable from the control unit. Replace in the reverse order of dismantling. Note: Make sure when the new cable is fitted and adjusted that the throttle slide does not stick when the handlebars are turned from side to side.

Replacement of the CD100 and CB125S cable

3 The twist grip on these models is of the Bendix type which operates on a simple spiral. Half of the spiral within the twist grip sleeve fits over a block which runs in a slot in the handlebar.
4 To remove the cable, unscrew the carburettor top and remove the cable from the slide, then undo the clamping screws which hold the right hand twist grip control to the handlebar. Twist the twist grip off; this will spiral off in the usual direction. Lift out the block with the nipple in it and then the block into which the outer cable is slotted. Take out the old cable and refit the new in reverse order of dismantling. Take care to ensure the throttle does not stick or the handlebars do not trap the cable on either lock.

9 Air cleaner - removal and replacement

1 Remove the right hand side cover by unclipping it from the frame.
2 Loosen the air intake at the carburettor mouth and remove the two nuts which locate and hold the unit.
3 Wash the foam element in a solvent, then wring it dry and apply a small amount of clean engine oil. Wring it out again. Replace the element, refit the element case and retighten the clamp. Replace the cover.
Warning: Do not use petrol (gasoline) to wash the filter element as it will dissolve the foam.

Fig. 2.1 Carburettor (Keihin)

1 Cover
2 Cable adjuster
3 Carburettor top
4 Gasket
5 Return spring
6 Needle retainer
7 Clip
8 Jet needle
9 Throttle slide
10 O-ring
11 Gasket
12 Throttle stop screw (tickover)
13 Pilot screw (mixture)
14 Needle jet
15 Needle jet holder
16 Main jet
17 Pilot jet
18 Sealing washer
19 Float needle seat
20 Float needle valve
21 Float
22 Pivot pin
23 Float bowl
24 Spring clip
25 Drain screw
26 Sealing washer

Chapter 2/Fuel system and lubrication

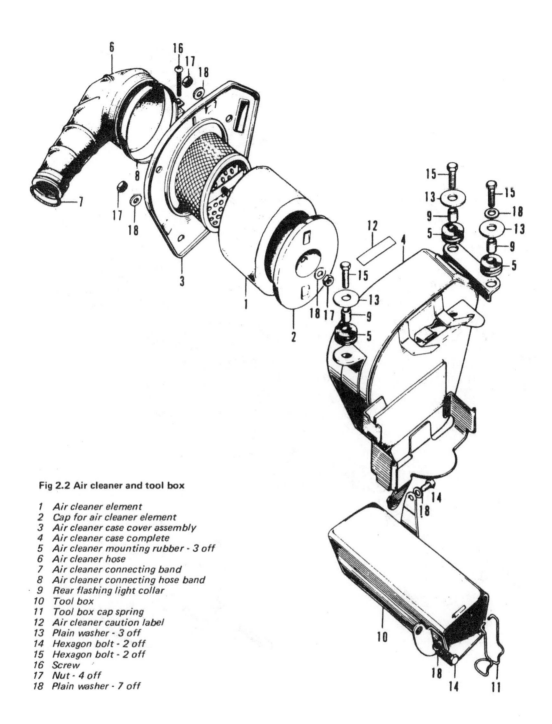

Fig 2.2 Air cleaner and tool box

1. Air cleaner element
2. Cap for air cleaner element
3. Air cleaner case cover assembly
4. Air cleaner case complete
5. Air cleaner mounting rubber - 3 off
6. Air cleaner hose
7. Air cleaner connecting band
8. Air cleaner connecting hose band
9. Rear flashing light collar
10. Tool box
11. Tool box cap spring
12. Air cleaner caution label
13. Plain washer - 3 off
14. Hexagon bolt - 2 off
15. Hexagon bolt - 2 off
16. Screw
17. Nut - 4 off
18. Plain washer - 7 off

10 Exhaust system - general

The exhaust system should not be altered in any way, such as removing the baffles to make more noise, as it does not necessarily follow that the more noise made the faster the machine. Racing machines have engines designed to go fast and noise is of little account. Tampering with the standard system designed to give optimum performance with as little noise as possible will upset the balance and cause reduced performance, even though the changed exhaust note creates the illusion of speed.

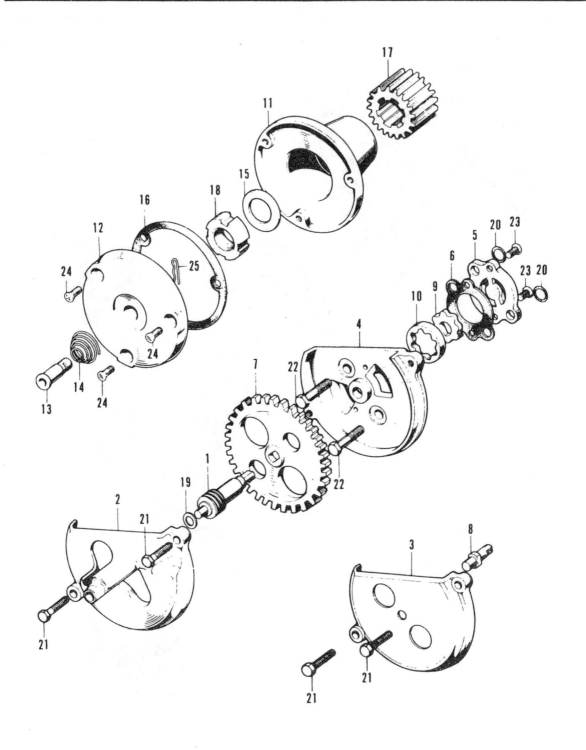

Fig 2.3 Oil pump and centrifugal oil filter

1. Tachometer drive pinion (39 teeth)
2. Oil pump pinion cover
3. Oil pump pinion cover
4. Oil pump body
5. Oil pump plate
6. Oil pump gasket
7. Oil pump drive pinion
8. Oil pump spindle
9. Oil pump inner rotor
10. Oil pump outer rotor
11. Oil filter rotor complete
12. Oil filter rotor cap
13. Pressure release valve
14. Spring for pressure release valve
15. Sleeve nut lock washer
16. Gasket for rotor cap
17. Primary drive pinion (18 teeth)
18. Sleeve nut
19. Thrust washer
20. O-ring - 2 off
21. Hexagon bolt - 2 off
22. Hexagon bolt - 2 off
23. Countersunk screw - 2 off
24. Cross-head screw - 3 off
25. Locking pin

12.3a Spring holds in gauze filter

12.3b Clean with solvent then replace

11 Engine lubrication - removal and replacement of the oil pump

1 For removal of the oil pump refer to Chapter 1.7.
2 When the oil pump is apart, check for signs of wear. This will be evident if the engine has at any time run low or out of oil. The wear will show as marks or deep pits; if in any doubt about any part, renew it without question. A further oil pump failure will wreck the whole engine.

12 Locating and cleaning the oil filters

1 The engine has two oil filters, one of the centrigual type and one gauze. The procedure for removing and cleaning each is to be found in Chapter 1.7. There is no necessity to remove the left hand side cover for access to the latter.
2 Once the cover is off the centrifugal filter scrape out all foreign bodies and wash out with petrol, dry and reassemble.
3 The gauze filter is in the lower exterior of the engine unit and should be cleaned with solvent and reassembled

Fault diagnosis chart overleaf

13 Fault diagnosis - fuel system and lubrication

Symptom	Cause	Remedy
Excessive fuel consumption	Air cleaner choked or restricted	Clean or renew.
	Fuel leaking from carburettor. Float sticking	Check all unions and gaskets. Float needle seat needs cleaning.
	Badly worn or distorted carburettor	Replace.
	Jet needle setting too high	Adjust as figure given in Specifications.
	Main jet too large or loose	Fit correct jet or tighten if necessary.
	Carburettor flooding	Check float valve and replace if worn.
Idling speed too high	Throttle stop screw in too far. Carburettor top loose	Adjust screw. Tighten top.
	Pilot jet incorrectly adjusted	Refer to relevant paragraph in this Chapter.
	Throttle cable sticking	Disconnect and lubricate or replace.
Engine dies after running for a short while	Blocked air hole in filler cap	Clean.
	Dirt or water in carburettor	Remove and clean out.
General lack of performance	Weak mixture; float needle stuck in seat	Remove float chamber or float and clean.
	Air leak at carburettor joint	Check joint to eliminate leakage, and fit new O ring.
Engine does not respond to throttle	Throttle cable sticking	See above.
	Petrol octane rating too low	Use higher grade (star rating) petrol.
Engine runs hot and is noisy	Lubrication failure	Stop engine immediately and investigate cause. Slacken cylinder head nut to check oil circulation. Do not restart until cause is found and rectified.

Chapter 3 Ignition system

Contents

General description ... 1	Auto-advance unit - removal, examination and replacement 5
Ignition coil - checking ... 2	Spark plug - checking and resetting the gap ... 6
Contact breaker - adjustment and timing the ignition ... 3	Fault diagnosis - ignition system ... 7
Condenser - removal and replacement ... 4	

Specifications

Spark plug

	NGK	Nippon Denso	Champion	KLG	Lodge
Make ...					
Type ...	D-8ES	X24ES	R6	TW270	HB12
Gap ...			0.024 - 0.028 in		
Gap ...			0.6 - 0.7 mm		
Reach ...			¾ in		

1 General description

The ignition system is extremely simple, comprising a battery, a set of points, a coil and condenser. As long as the spark plug is kept in good condition, the points are clean and have the correct gap and the timing is right, the system will generally give no trouble.

2.1 Ignition coil and condenser are located above the head steady, under the petrol tank

2 Ignition coil - checking

1 The ignition coil is a sealed unit and designed to give long service without need for attention. The coil is located under the petrol tank, which must be removed to gain access.
2 To check whether the coil is defective, disconnect the spark plug lead from the cap and hold the lead about 3/16 inch or roughly 6 mm away from the cylinder head. Turn on the ignition and flick the points open with an insulated screwdriver. This should produce a good blue spark between the lead and the cylinder head.
3 If the spark is weak or intermittent, a check will have to be carried out by a Honda dealer as this required specialist equipment.

3 Contact breaker - adjustment and timing the ignition

1 To set the points gap it is necessary to take off the points cover and the lower engine cover over the rotor, on the left side of the machine.
2 With both covers removed, rotate the engine until the 'F' mark on the rotor lines up with the marker; move forward slowly until the points are fully open.
3 First check that the points are clean and that no pitting has occurred. If they are finely pitted, clean with fine abrasive paper or a points file, but if they are badly pitted, fit a new set of points. To adjust the gap, insert a 0.012 inch feeler gauge into the gap and slacken the clamping screw. Adjust until a slight drag is felt on the feeler gauge blade, then retighten. Recheck after tightening because the setting may vary during this operation.
4 With the points gap set, use a timing light (or Stroboscope) if available, to make a final check. Take a 0.0015 inch feeler gauge and turn the contact breaker backplate so that when the 'F' mark lines up with the pointer the contact breaker points

3.2 "F" mark on rotor in line

are just starting to open. With the feeler gauge method maintain a very light pull on the gauge whilst slowly rotating the engine. When the 'F' mark lines up, the feeler gauge should slide out, assuming the setting is correct.

5 Connect the timing light and start the engine. The two marks on the rotor should swing around and line up with the pointer when the engine is revved. Ideally the two marks should hover one each side of the pointer. There may be a small amount of oil thrown out when adopting this method so do not get too close.

4 Condenser - removal and replacement

1 The condenser is located under the tank, adjacent to the coil. If the machine becomes difficult to start or misfiring occurs, then it is possible that the condenser is at fault.

2 Examine the contact breaker points whilst the engine is running to see if arcing is occurring. When the engine has stopped, check for damage to the faces of the points in the form of blackness or burning. This is characteristic of condenser failure.

3 Sophisticated test equipment is required for the condenser to be checked, so it is better to fit a new one and observe the effect on the engine performance in view of the relatively low cost of this component.

5 Auto-advance unit - removal, examination and replacement

1 Fixed ignition timing is of little advantage as the engine speed increases, and it is necessary to incorporate a method of advancing the timing by centrifugal means. A balance weight assembly located behind the points, linked to the cam, is used on the Honda singles. A centre fixing bolt holds the unit secure, permitting the cam to move by means of this linkage and so advance the ignition timing.

2 Remove the contact breaker points cover and also the backplate. Unscrew the centre fixing bolt and pull off the unit.

3 Inspect the unit to see if the springs are broken or if any wear is evident. This unit can also suffer from rust due to condensation which is evident if the cam will not move freely. If any malfunction or breakage has occurred, renew the complete unit, but if it appears to be in good condition, lightly oil it and slide it back over the end of the camshaft, checking the slot locates correctly with the pin on the camshaft. Replace the centre bolt, tighten and continue reassembly in reverse order.

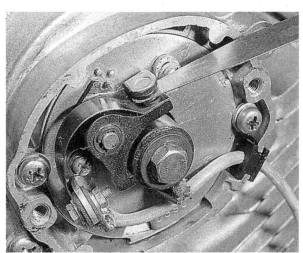

3.3 Feeler gauge inserted into points gap

3.4 Slacken clamping screws to rotate backplate

5.3 The advance and retard unit in location

Electrode gap check - use a wire type gauge for best results

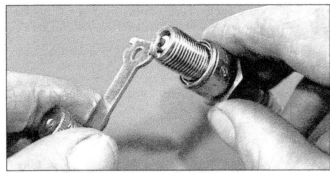

Electrode gap adjustment - bend the side electrode using the correct tool

Normal condition - A brown, tan or grey firing end indicates that the engine is in good condition and that the plug type is correct

Ash deposits - Light brown deposits encrusted on the electrodes and insulator, leading to misfire and hesitation. Caused by excessive amounts of oil in the combustion chamber or poor quality fuel/oil

Carbon fouling - Dry, black sooty deposits leading to misfire and weak spark. Caused by an over-rich fuel/air mixture, faulty choke operation or blocked air filter

Oil fouling - Wet oily deposits leading to misfire and weak spark. Caused by oil leakage past piston rings or valve guides (4-stroke engine), or excess lubricant (2-stroke engine)

Overheating - A blistered white insulator and glazed electrodes. Caused by ignition system fault, incorrect fuel, or cooling system fault

Worn plug - Worn electrodes will cause poor starting in damp or cold weather and will also waste fuel

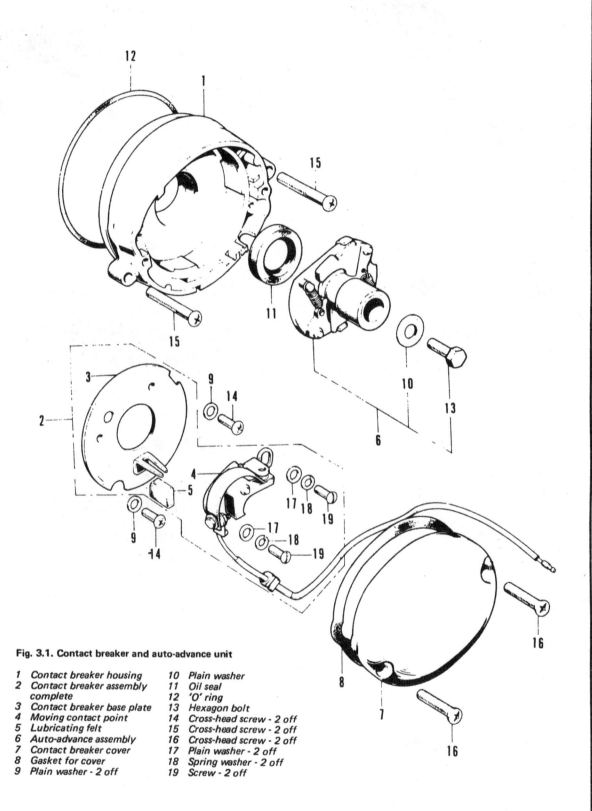

Fig. 3.1. Contact breaker and auto-advance unit

1 Contact breaker housing
2 Contact breaker assembly complete
3 Contact breaker base plate
4 Moving contact point
5 Lubricating felt
6 Auto-advance assembly
7 Contact breaker cover
8 Gasket for cover
9 Plain washer - 2 off
10 Plain washer
11 Oil seal
12 'O' ring
13 Hexagon bolt
14 Cross-head screw - 2 off
15 Cross-head screw - 2 off
16 Cross-head screw - 2 off
17 Plain washer - 2 off
18 Spring washer - 2 off
19 Screw - 2 off

Chapter 3/Ignition system

6 Spark plug - checking and resetting the gap

1 The engine uses as standard a spark plug of the correct heat range and hardness and it is not advisable to fit other plugs of varying heat range unless instructed by an authorised Honda dealer or agent.
2 To check the gap use a feeler gauge. The correct gap should be within the range 0.6 - 0.7 mm (0.025 - 0.028 inch). Insert the gauge into the gap and give a very light tap on the side electrode which projects over the centre electrode. If the gap is too large tap down until the feeler gauge is a good sliding fit.
3 The plug should be cleaned regularly and it is strongly advised that a new plug be carried in the tool box at all times. Always keep spare plugs clean and dry. To clean the plug, use a sand blaster or a wire brush. Clean the gap with fine emery paper and always check and reset the gap after cleaning.
4 When the plug has been cleaned and gapped, push the plug into the plug cap, turn on the ignition and rest the plug outer body firmly on the engine. Using the kickstarter, kick the machine over and watch the spark. If the spark jumps to the side of the plug it means that the plug has broken down and a new replacement is necessary. If all appears correct, replace the plug in the machine, tighten and replace the plug cap.

7 Fault diagnosis - ignition system

Symptom	Cause	Remedy
Engine will not start	Faulty ignition switch	Operate switch several times in case contacts are dirty. If lights and other electrics function, switch may need replacement.
	Short circuit in wiring	Check whether fuse is intact. Eliminate fault before switching on again.
Engine misfires	Faulty condenser in ignition circuit	Replace condenser and retest.
	Fouled spark plug	Replace plug and have original cleaned.
	Poor spark due to generator failure and discharging battery	Check output from generator. Remove and recharge battery.
Engine lacks power and overheats	Retarded ignition timing	Check timing and also contact breaker gap. Check whether auto-advance mechanism has jammed.
Engine 'fades' when under load	Pre-ignition	Check grade of plugs fitted; use recommended grades only. Verify whether lubrication system has pressure.

Chapter 4 Frame and forks

Contents

General description ... 1	Centre stand - examination and removal (with brake pedal) 13
Front forks - removal from frame ... 2	Footrest bar and prop stand - removal and replacement ... 14
Front forks - dismantling and examination ... 3	Speedometer head and tachometer head - removal and replacement ... 15
Fork yokes and steering head races - examination ... 4	Speedometer and tachometer cables - examination and renovation ... 16
Front forks - reassembly ... 5	Dualseat - removal ... 17
Front forks - refitting to frame ... 6	Petrol tank embellishments - removal and replacement ... 18
Frame assembly - examination and renovation ... 7	Steering lock - removal and replacement ... 19
Swinging arm - examination and renovation ... 8	Cleaning the machine - general ... 20
Rear suspension units - examination and renovation ... 9	Fault diagnosis - frame and forks ... 21
Rear suspension units - adjusting the loading ... 10	
Removing and replacing the side covers ... 11	
Prop stand - examination ... 12	

Specifications

Frame
Type:
- CB100, CL100, CB125S, CD125S models ... Open diamond
- SL100 and SL125 trail models ... Double tube cradle

Front suspension
Type:
- All models ... Telescopic forks
- Travel ... 83 mm (3.2 in)

Fork springs:
- Free length:
  - CB100 and CL100 models ... 184 mm (7.2440 in)
  - CB125S and CD125S models ... 205.5 mm (8.0905 in)
  - SL100 model ... 484.2 mm (19.0629 in)
  - SL125 model ... 482.3 mm (18.9881 in)
- Wear limit:
  - CB100 and CL100 models ... 160 mm (6.2992 in)
  - CB125S and CD125S models ... 180 mm (7.0866 in)
  - SL100 model ... 460 mm (18.1102 in)
  - SL125 model ... 460 mm (18.1102 in)

Oil content (dry):
- CB100, CD100 and CB125S models ... 130 - 140 cc
- SL100 and SL125 models ... 180 - 190 cc

Oil content (after draining):
- CB100, CL100 and CB125S models ... 110 - 120 cc
- SL100 and SL125 models ... 165 - 175 cc

Rear suspension units
Springs:
- Free length:
  - CB100, CL100, CB125S and CD125S models ... 180.9 mm (7.1200 in)
  - SL100 and SL125 models ... 190 mm (7.4803 in)
- Wear limit:
  - CB100, CL100, CB125S and CD125S models ... 160 mm (6.2992 in)
  - SL100 and SL125 models ... 170 mm (6.6929 in)

Chapter 4/Frame and forks

1 General description

The lightweight frame used on the Honda singles relies on the engine to bridge the gap between the front downtube and the main spine. The SL100 and SL125 trail models have a different frame of the double cradle type.

The front forks are oil damped as are the rear suspension units. They differ in that the front fork oil can be changed but the rear units are sealed. The forks fitted to the SL trail models have internal springs.

2 Front forks - removal from the frame

1 It is unlikely that the front forks will need to be removed from the frame as a complete unit unless the steering head bearings require attention or the forks are damaged in an accident.
2 Commence operations by placing the machine on its centre stand and disconnecting the front brake by slackening back the adjuster and removing it from the front brake lever.

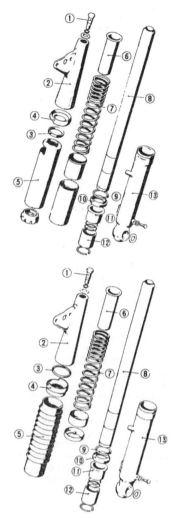

CB100 & CB125S/CL100 & CD125S
1 Front fork bolt
2 Front fork upper cover
3 Fork cover lower seat packing
4 Fork cover lower seat
5 Front fork under cover (CB100), front fork boot (CL100)
6 Front fork spring guide
7 Front fork spring
8 Stanchion
9 37 mm circlip
10 Front fork oil seal
11 Front fork bush
12 Front fork piston
13 Lower fork leg

SL100 & SL125
1 Front fork bolt
2 Front fork upper cover
3 Front fork dust seal
4 44 mm circlip
5 Back up ring
6 31 x 43 x 10 oil seal
7 Front fork bush
8 Front fork spring
9 Stanchion pipe
10 Front fork piston
11 Lower fork leg

Fig. 4.1. Front fork assemblies

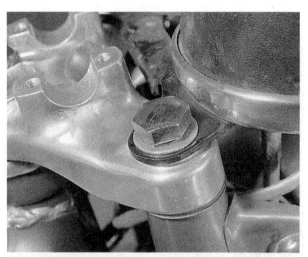

2.6 Remove the top stanchion bolts

2.7 Slacken the large chrome centre nut

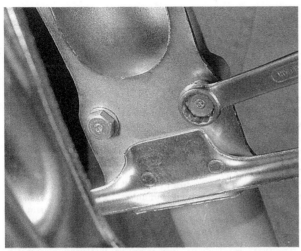

2.9 Take off the mudguard by removing the four fixing bolts

2.13 Slacken both pinch bolts

3.5 Remove circlip

3 It is not necessary to take the handlebars off the machine. Slacken the screw which retains the headlamp reflector unit and place the unit to one side. Do not be afraid to unplug any wires because they are all colour-coded and extremely easy to plug back together. Follow the wires from the handlebars through to the headlamp shell and unplug them all.
4 Next remove the four 6 mm bolts which secure the clamps that hold the handlebars. Take off the clamps, place some rag on the petrol tank and rest the handlebars on the tank.
5 Unfasten the mounting bolt for the speedometer, disconnect the drive cable and unplug the wire within the headlamp shell which comes from the speedometer lamp.
6 Suspend the front of the machine, taking the weight off the front wheel by blocking the bike up. Remove the large bolts on top of the fork stanchions.
7 Next slacken the chrome nut which holds the top yoke to the steering stem. Do not take it off yet.
8 Straighten out the split pin and undo the nut on the end of the front wheel spindle. Pull out the spindle and remove the wheel.
9 Remove the bolts from the inside of the fork legs which retain the mudguard in position.
10 Place a drip tray under each fork leg and remove the drain plug found on the outside of the fork legs, near the bottom, and allow all the oil to drain out.
11 Take out the bolts supporting the headlamp and remove the headlamp shell.
12 Remove the large chrome nut slackened earlier, and remove the top yoke. Now it is possible to move the handlebar further back because the throttle cable is free. Pull the wiring harness out of the way and tuck it in beside the petrol tank.
13 Slacken the bolts on the bottom yoke clamping the fork stanchions in place and, one at a time, ease each out in a downwards direction. Be careful to note how the various covers come off and to catch them as they become free.
14 With both the fork legs out of the yokes, support the bottom yoke with one hand and undo the top nut with the other. Do not let go when it is off as there are a large number of ball bearings in the steering head races. (Twenty-one in each set.) Wrap some rag around the lower bearing, take out the bottom yoke and collect the ball bearings that will be released as the cup and cone of the bearing separate. Most probably the ball bearings in the upper race will remain in place.

3 Front forks - dismantling and examination

CB100 – CB125S and CL100 – CD125S
1 There is not much left to complete the strip of the individual fork legs as the springs are external; only the bushes are located internally.

SL100 – SL125
2 On these models the fork springs are internal, but the stripdown procedure is very similar.
3 Both types of fork are stripped using virtually a similar technique. For ease of explanation the CB100, CB125S, CL100 and CD125S models will be referred to as the CB125S and the SL100 to SL125 will be referred to as the SL125.
4 On the SL125 pull off the rubber cap which fits around the top of the slider and use circlip pliers to extract the large circlip which holds in the oil seal and the top fork bush.
5 On the CB125S tap off the top shroud or gaiter using a soft alloy drift and use circlip pliers to release the circlip.
6 To remove the bushes from either type of fork assembly, hold the bottom of the leg in a vice fitted with soft jaws so that the fork leg does not get damaged. Then catching hold of the fork stanchion, slide it vigorously in and out until the assembly is free and can be lifted out complete with bushes. Repeat with the other fork leg.
7 One of the first items to look for are split, cut or perished oil seals. This would usually be evident before stripping, as oil would

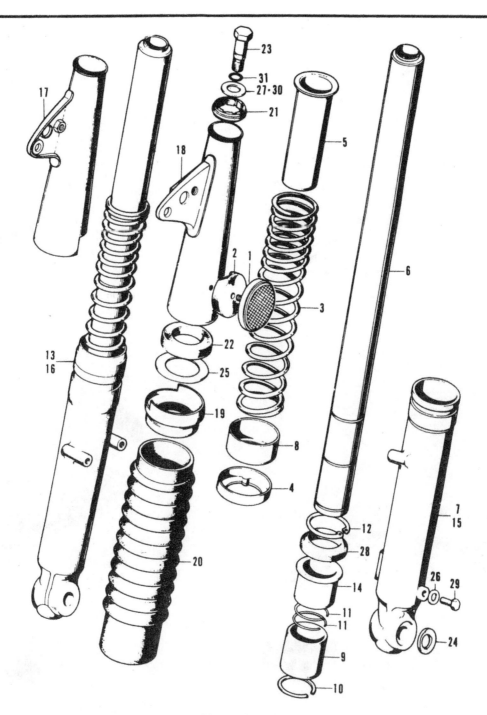

Fig. 4.2. Front forks

1	Reflector - 2 off	17	Right-hand upper fork shroud
2	Reflector base - 2 off	18	Left-hand upper fork shroud
3	Fork spring - 2 off	19	Seating for shroud - 2 off
4	Fork spring seating - 2 off	20	Fork gaiter - 2 off
5	Fork spring guide - 2 off	21	Fork shroud upper mounting - 2 off
6	Stanchion - 2 off	22	Fork shroud lower mounting - 2 off
7	Lower fork leg - left-hand	23	Front fork bolt - 2 off
8	Fork spring seating guide - 2 off	24	Washer - 2 off
9	Damper piston - 2 off (acts also as lower fork bush)	25	Washer for fork shroud - 2 off
10	Piston snap ring - 2 off	26	Drain plug washer - 2 off
11	Piston ring stop ring - 2 off	27	Thrust washer - 2 off
12	Internal circlip - 2 off	28	Oil seal - 2 off
13	Right-hand fork leg complete	29	Hexagon bolt - 2 off (drain plug)
14	Upper fork bush - 2 off	30	Washer for fork bolt - 2 off
15	Lower fork leg complete - left-hand	31	'O' ring
16	Left-hand fork leg complete		

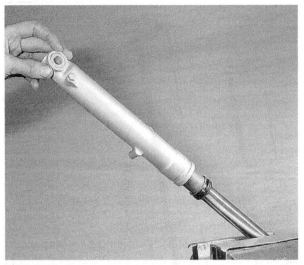
3.6 Pull sharply to free fork leg from stanchion

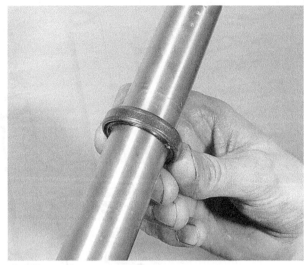
3.7 Check oil seals for damage

3.8a Check top bush for wear

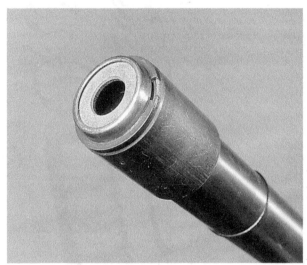
3.8b Check bottom bush for wear

4.2a Examine cups for marks, dents or cracks

4.2b To renew tap out with a drift

seep out over the slider. Replacement of the fork oil seal is a matter of course as they are reasonably priced.

8 Broken or compressed springs, slack in the bushes or oil leaks are other points to look for. The bushes should be renewed if they are worn and the free length of the springs checked against the figures given in the Specifications Section of this Chapter. If they are outside the serviceable limit, they must be renewed. Never replace a single spring, always in pairs.

9 Check the fork legs for damage if the machine has been in an accident, especially for cracks or breakage. A repair by welding is impracticable. Always replace with a good secondhand or new part.

10 The fork stanchions must be checked for straightness with the aid of V blocks. They can be straightened if less than 3/16 inch out of true over the total length, but if more than this, they should be scrapped and replaced.

4 Fork yokes and steering head races - examination

1 To check the top yoke for accident damage, push the fork stanchions through the bottom yoke and fit the top yoke. If it lines up, it can be assumed the yokes are not bent. Both must also be checked for cracks. If they are damaged or cracked, fit new replacements.

2 To check the steering head races, inspect all of the ball bearings for pits or cracks and also, if the machine has been involved in an accident, the race cups which are in the frame. They, more often than not, become indented by the ball bearings which gives a 'lumpy' effect when riding. If there is any doubt about the condition of the head races, renew the whole set without question.

5 Front forks - reassembly

1 Fit new bushes to the fork stanchions by easing off the bottom circlip, pulling off the old bush and fitting the new bush in its place. Replace the circlip. Slide the old upper bush up the stanchion after first removing the oil seal. Slide the new bush into position with the flange uppermost, followed by the oil seal.

2 Oil liberally, then insert the stanchion into the fork leg, tap down the seal and replace the circlip.

6 Front forks - refitting to the frame

1 If it has been necessary to remove the fork assembly completely from the frame, refitting is accomplished by following the dismantling procedure in reverse. Check that none of the balls are misplaced whilst the steering head stem is passed through the head set. It has been known for a ball to be displaced, drop down and wear a groove or even jam the steering, so be extremely careful in this respect.

2 Take particular care when adjusting the steering head bearings; the adjusting nut should be locked in place with the top yoke nut sufficiently to remove all the slack, but not tight otherwise damage to the cups or ball bearings will occur. When adjusted, lightly flick the handlebars with the finger. They should drop easily from side to side, but if the fork legs are held and pulled forwards and backwards no play should be felt.

3 Difficulty may be experienced in raising the fork stanchions so that the end locates in the top yoke. If a stanchion puller is not available use a piece of wooden dowel such as a broom handle, force it in the thread and use it to pull the stanchion into place.

4 Before refitting the top fork bolts do not forget to replace the oil; the quantity required is shown in the Specifications Section of this Chapter. Do not forget to tighten the drain plug first. Note that the fork leg with the brake plate location fits on the left hand side of the machine. On the CB125S series attach the lower shroud and the headlamp bracket whilst sliding the stanchion into place.

5.1 Remove bottom circlip and off comes bottom bush

6.1a Assemble ball bearings in bottom race with grease

6.1b Build up top race in the same way

6.1c Insert upper cone and screw on adjuster nut

6.1d Fit top yoke but do not adjust at this stage

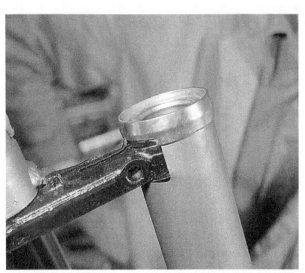

6.3a Before replacing stanchions on 125S etc., replace the bottom cover over bottom yoke

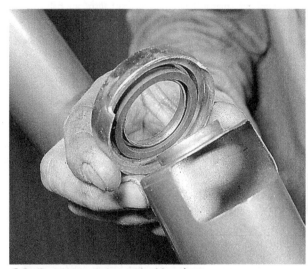

6.3b Remember spacers and rubber rings

6.4a Place external springs over stanchion before insertion

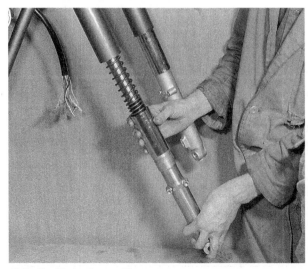

6.4b Feed in both stanchions

Chapter 4/Frame and forks

6.4c Refill with correct quantity of oil

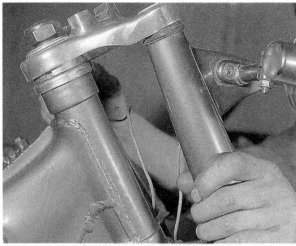

6.5 Line up headlamp brackets before tightening top bolts

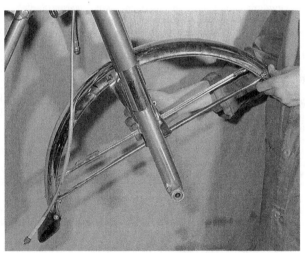

6.6a Make sure cable clip is on left-hand side

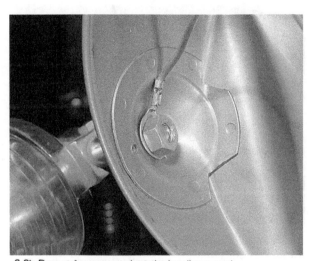

6.6b Do not forget to replace the headlamp earth

7.1 Ideal way to check frame

5   Tighten the top mounting bolts and then the clamp bolts in the bottom yoke.

6   Replace the front mudguard, headlamp, speedometer, rev counter, handlebars and tighten. Reconnect the cables, wiring and any other extras.

7   When replacing the front wheel, first make sure that the speedometer drive pegs inside the front hub have not ridden up out of their location. After checking this, insert the wheel spindle (not forgetting the spacer on the opposite side to the brake). Do not fail to check the brake location. The peg aligns with the slot on the outside of the brake plate. Push through the spindle and place on the washer and the nut. Tighten the nut and fit a new split pin. Fit and adjust the brake cable, also the speedometer drive cable.

7 Frame assembly - examination and renovation

1   If the machine is stripped for a complete overhaul, this affords a good opportunity to inspect the frame for cracks or other damage which may have occurred in service. Check the front downtube immediately behind the steering head and the top tube immediately behind the steering head, the two points where fractures are most likely to occur. The straightness of the tubes

8.3 Remove both suspension unit retaining nuts

8.4 Swinging arm pivot nut

concerned will show whether the machine has been involved in a previous accident.

2 If the frame is broken or bent, professional attention is required. Repairs of this nature should be entrusted to a competent repair specialist, who will have available all the necessary jigs and mandrels to preserve correct alignment. Repair work of this nature can prove expensive and it is always worthwhile checking whether a good replacement frame of identical type can be obtained from a breaker or through any form of Service Exchange Scheme. The latter course of action is preferable because there is no safe means of assessing on the spot whether a secondhand frame is accident damaged too.

8 Swinging arm - examination and renovation

1 To check for wear in the swinging arm bushes, hold the frame firmly in one hand and with the other hold the rear wheel. There should not be any side play at all; if there is, replacement of the swinging arm pivot bearings may be necessary.

2 Remove the rear wheel assembly as described in Chapter 5.6 and disconnect the chain, but do not pull it right off. Leave on the gearbox sprocket.

3 Remove the two rear suspension units from their upper and lower locating points and then remove the chainguard.

4 Undo the large nut on the end of the swinging arm spindle and pull out the pin whilst supporting the swinging arm in the other hand.

5 To replace the bushes a press is necessary with a few good fitting pieces of short tube, but a vice clamped rigidly to a workbench will often suffice.

6 Position one piece of tube to support the swinging arm and line up another against the bush that has the same outside diameter in a vice fitted with soft jaws. Close the vice, using the tube to press the old bush out of position. Use the same method to press the new bush into position and when both have been renewed, reassemble in the reverse order of dismantling. Do not forget to replace and retighten the rear brake torque arm.

9 Rear suspension units - examination and renovation

1 Remove both the rear units from their top and bottom locations and clean them off.

2 Clamp firmly (without distorting) the lower part of the unit in a vice and place a piece of bar or a strong screwdriver blade through the lower unit mounting lug. Unscrew and take off the lower mounting.

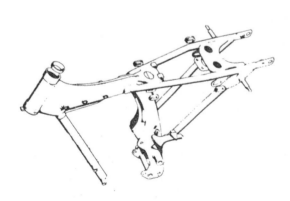

Fig. 4.3A. CB100, CL100, CB125S, CD125S 'Diamond' frame

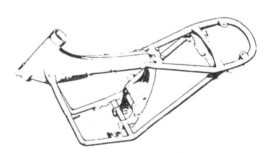

Fig. 4.3B. SL100, SL125 Double cradle frame

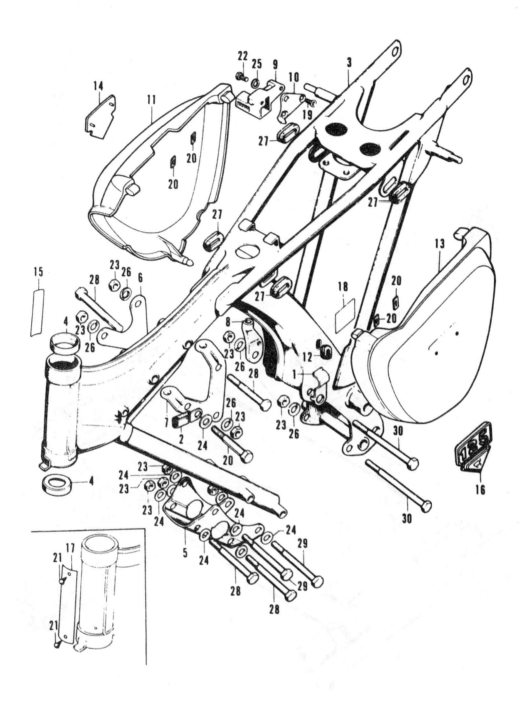

Fig. 4.4. Frame and side covers (except SL models)

1. Generator cable clip
2. Ignition/lighting switch cable clip
3. Frame complete
4. Steering head cup - 2 off
5. Front engine plate
6. Right-hand head steady plate
7. Left-hand head steady plate
8. Clutch cable guide
9. Helmet holder
10. Helmet holder stay
11. Right-hand side cover
12. Air cleaner case grommet
13. Left-hand side cover
14. Right-hand side cover emblem
15. Nameplate
16. Left-hand side cover emblem
17. Registration number plate
18. Battery caution transfer
19. Flat head screw - 2 off
20. Spire nut - 4 off
21. Rivet screw - 2 off
22. Hexagon bolt
23. Nut - 9 off
24. Plain washer - 9 off
25. Spring washer
26. Spring washer - 5 off
27. Mounting rubber - 4 off
28. Engine bolt - 5 off
29. Engine bolt - 2 off
30. Rear engine bolt - 2 off

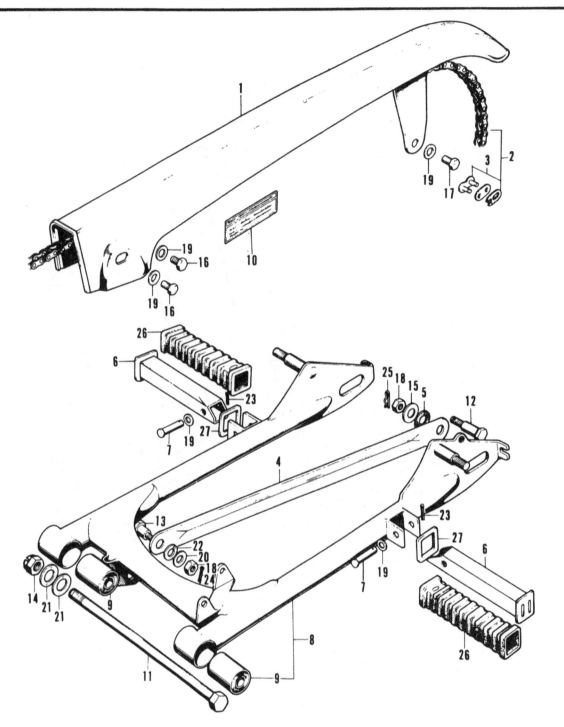

Fig. 4.5. Swinging arm fork and chainguard

1. Chainguard
2. Final drive chain
3. Spring link
4. Rear brake torque arm
5. Rubber bush for torque arm anchorage
6. Pillion footrest - 2 off
7. Pivot pin for pillion footrest - 2 off
8. Swinging arm fork complete
9. Rubber bush for swinging arm fork - 2 off
10. Tyre caution transfer
11. Pivot bolt for swinging arm fork
12. Rear brake plate torque arm bolt
13. Rear brake torque arm anchor bolt
14. Self-locking nut for pivot bolt
15. Plain washer
16. Hexagon bolt - 2 off
17. Hexagon bolt
18. Nut - 2 off
19. Plain washer - 5 off
20. Plain washer
21. Plain washer - 2 off
22. Spring washer
23. Split pin - 2 off
24. Split pin
25. Locking pin
26. Pillion footrest rubber - 2 off
27. Pillion footrest washer - 2 off

Chapter 4/Frame and forks

3 On the SL100 and SL125 models the rear suspension spring has no covers but is adjustable for spring loading.

4 Once the units have been stripped check each shock absorber unit for signs of an oil leak. If a leak or seepage is found, the damper unit must be renewed as it is a sealed unit and cannot be repaired. Check the springs for length (they should be equal) and cracks. Also check the shock absorber action when pulled; it should move in and out quite freely when moved very slowly but become increasingly difficult to move as the action is speeded up. If the action does not get more difficult, it is safe to assume that the unit is not functioning correctly and should be renewed.

5 After inspection, reassemble and refit in reverse order of dismantling. If difficulty occurs, refer to the diagrams, one for the SL series and the other for the CB, CL and CD models.

10 Rear suspension units - adjusting the loading

SL100 and SL125 only

1 As mentioned earlier in the text, the units can be adjusted without detaching them from the machine, so that the loading can be matched to that most suitable for the conditions under which the machine is to be used. A built-in cam at the lower end of each unit permits the sleeve carrying the lower end of the compression springs to be rotated, so that the sleeve is adjustable to different heights. The lowest position, or lightest loading is recommended for solo riding, the middle or medium loading for a heavy solo rider or one with luggage to carry, and the highest or heaviest loading when a pillion passenger is carried. A special 'C' spanner, provided with the tool kit, is used to effect the adjustments.

2 Both units must always be set to an identical rating, otherwise the handling of the machine will be seriously affected.

11 Removing and replacing the side covers

1 It is an extremely easy task to remove either the left or right hand panel. Start by pulling the side panel at the back until it unclips, then pull from the bottom point until that too comes unclipped. Finally, pull at the front, when the cover will be fully released.

2 If difficulty is found in getting the cover back on (or off for that matter) smear a little soap on the three rubbers captive in the frame; line up the three mounting spigots and hit firmly with the flat of the hand.

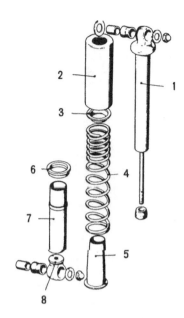

Fig. 4.6A. Rear suspension unit - CB100, CL100, CB125S, CD125S

1 Rear shock absorber
2 Rear shock absorber upper shroud
3 Rear shock absorber upper seat
4 Rear shock absorber spring
5 Rear shock absorber spring guide
6 Rear shock absorber spring seat
7 Rear shock absorber lower leg
8 Rear shock absorber bottom mounting

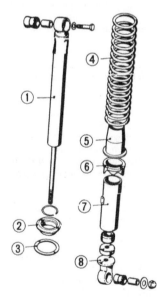

Fig. 4.6B. Rear suspension unit - SL100, SL125

1 Rear shock absorber
2 Rear shock absorber spring seat stopper
3 Rear shock absorber spring upper seat
4 Rear shock absorber spring
5 Rear shock absorber spring guide
6 Rear shock absorber spring adjuster
7 Rear shock absorber lower leg
8 Rear shock absorber bottom mounting

12.1 Removal of this split pin releases footbrake and stand

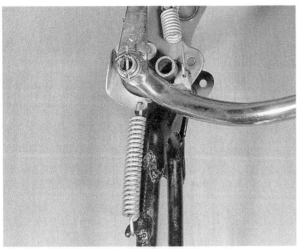

12.4 Centre stand and footrest lever in position on bare frame

12 Prop stand - examination

1 A prop stand which pivots from the left hand lower frame tube is provided for occasional parking, when it is not considered necessary to use the centre stand.
2 At regular intervals check that the prop stand return spring is in good order and that the pivot is lubricated.

13 Centre stand - examination and removal (with brake pedal)

1 The centre stand is attached to the lower rear part of the frame by the same tubular spigot as the foot brake.
2 To remove the stand, or rear brake pedal, straighten out the split pin on the right side of the machine and pull it out. If the spigot is badly worn, renew it otherwise grease and replace it, using a new split pin.

14 Footrest bar and prop stand - removal and replacement

Place the machine on the centre stand and on the underside of the engine will be found four bolts holding the footrest bar and prop stand to the machine. Remove these four bolts and drop the left hand side down, wriggling it out from between the silencer and the foot brake on the right. To replace the assembly, reverse the order of dismantling and insert and tighten firmly the four bolts. Check these from time to time as the side stand takes the weight of the machine when it is in use.

15 Speedometer head and tachometer head - removal and replacement

1 The speedometer (and tachometer on some models) are secured to a bracket that bolts to the top fork yoke, in which the speedometer head (and tachometer head) are rubber mounted to prevent internal damage by vibration.
2 To remove either instrument unscrew the drive cable and take off the headlamp unit to allow the indicator and internal lighting lamps to be disconnected. Unscrew the mounting nuts and remove. Replace in reverse order of dismantling. Wire reconnection is easy due to the colour-coded system used.
3 Apart from defects in either the drive or the cable, a speedometer or tachometer that malfunctions is difficult to repair. Fit a replacement or alternatively, obtain one from a crash repair specialist who sometimes have perfectly good instruments with scratched cases and glass etc. Remember that a speedometer in good working order is a statutory requirement.

16 Speedometer and tachometer cables - examination and renovation

1 It is advisable to detach the speedometer and tachometer drive cables from time to time in order to check whether they are adequately lubricated and whether the outer covers are compressed or damaged at any point along their run. A jerky or sluggish movement at the instrument head can often be attributed to a cable fault.
2 To grease the cable, uncouple both ends and withdraw the inner cable. After removing the old grease, clean with a petrol-soaked rag and examine the cable for broken strands or other damage.
3 Regrease the cable with high melting point grease, taking care not to grease the last six inches closest to the instrument head. If this precaution is not observed, grease will work into the instrument and immobilise the sensitive movement.
4 If the cable breaks, it is usually possible to renew the inner cable alone, provided the outer cable is not damaged or compressed at any point along its run. Before inserting the new inner

15.1 Tachometer drive located at front of engine

15.2 Speedometer drive on front wheel

Chapter 4/Frame and forks

18.2 To remove unscrew the two countersunk screws

cable, it should be greased in accordance with the instructions given in the preceding paragraph. Try and avoid tight bends in the run of a cable because this will accelerate wear and make the instrument movement sluggish.

17 Dualseat - removal

1 The dualseat is fixed to the rear of the machine by two brackets bolted to the rear part of the frame assembly. At the front there is a bracket under the seat which engages with a location bracket on the frame just to the rear of the petrol tank.
2 To remove the seat, remove the two fixing bolts. Hold the seat firmly at the rear, pull gently and the seat will then become detached. Replace by reversing this procedure.

18 Petrol tank embellishments - removal and replacement

CB, CD, CL models only
1 The tank badges are held in place by small spring clips which clip into small holes set into the tank side.
2 To remove the badge a great deal of patience is needed. Put two layers of plastic tape all round the badge to protect the tank, then using a thin, strong object, work a little at one end and then at the other to lever the badge off. If possible, renew the clips when reassembling.
3 Removing and replacing the tank badge is a tricky job since the badges are easily broken. The only time when they need to be removed is during a re-spray or when crash damage has occurred.

19 Steering lock - removal and replacement

1 The steering lock is situated at the front of the machine, by the bottom fork yoke. It is advisable to use it whenever the machine is left unattended, even for a short time.
2 To remove the lock, remove the countersunk screws holding the lock body to the bottom yoke. On some models the forks do not have to be removed from the frame before this action can be taken. Replace and reassemble in reverse order. Remember you will have to carry an additional key for the new lock.

20 Cleaning the machine - general

1 After removing all surface dirt with a rag or sponge washed frequently in clean water, the application of car polish or wax will give a good finish to the machine. The plated parts should require only a wipe over with a damp rag, followed by polishing with a dry rag. If, however, corrosion has taken place, which may occur when the roads are salted during the winter, a proprietary chrome cleaner can be used.
2 The polished alloy parts will lose their sheen and oxidise slowly if they are not polished regularly. The sparing use of metal polish or special polish such as Solvol Autosol will restore the original finish with only a few minutes labour.
3 The machine should be wiped over immediately after it has been used in the wet so that it is not garaged under damp conditions which will cause rusting and corrosion. Make sure the chain is wipeed and if necessary re-oil it to prevent water from entering the rollers and causing harshness with an accompanying rapid rate of wear. Remember there is little chance of water entering the control cables if they are lubricated regularly, as recommended in the Routine Maintenance Section.

Fault diagnosis chart overleaf

21 Fault diagnosis - frame and forks

Symptom	Cause	Remedy
Machine veers to left or right with hands off handlebars	Incorrect wheel alignment Bent forks Twisted frame	Check and re-align. Check and replace. Check and replace.
Machine rolls at low speeds	Overtight steering head bearings	Slacken and re-test.
Machine judders when front brake is applied	Slack steering head bearings	Tighten until all play is taken up.
Machine pitches badly on uneven surfaces	Ineffective fork dampers Ineffective rear suspension units	Check oil content. Check damping action.
Fork action stiff	Fork legs out of alignment (twisted in yokes)	Slacken yoke clamps, front wheel spindle and fork top bolts. Pump forks several times then tighten from bottom upwards.
Machine wanders. Steering imprecise, rear wheel tends to hop	Worn swinging arm pivot	Dismantle and replace bushes and pivot shaft.

Chapter 5 Wheels, brakes and tyres

Contents

General description ... 1	Rear wheel sprocket - removal and replacement ... 9
Front wheel - examination and removal ... 2	Rear wheel - replacement ... 10
Front brake assembly - examination, renovation and reassembly ... 3	Final drive chain - examination and lubrication ... 11
	Final drive chain - adjustment ... 12
Front wheel bearings - removal and replacement ... 4	Front and rear brake adjustment ... 13
Front wheel - replacement ... 5	Front wheel balancing ... 14
Rear wheel - examination and removal ... 6	Tyres - removal and replacement ... 15
Rear wheel bearings - removal and replacement ... 7	Security bolts ... 16
Rear brake - removal, examination and replacement ... 8	Fault diagnosis - wheel, brakes and tyres ... 17

Specifications

Brakes
- Inside diameter of brake drums ... 109.8 - 110.2 mm (4.3229 - 4.3385 in)
- Wear limit ... 112 mm (4.4094 in)

Brake linings
- Lining thickness ... 3.9 - 4.1 mm (0.1535 - 0.1614 in)
- Wear limit ... 2 mm (0.0787 in)
- Swept area of drums ... 86.4 cm^2 (13.4 sq in)

Tyres
Size:
Front:
- CB100 and CD125S models ... 2.50 x 18 inch
- SL100 model ... 2.75 x 19 inch
- CB125S model ... 2.75 x 18 inch
- SL125 model ... 2.75 x 21 inch
- CD 125S model ... 2.50 x 18 inch

Rear:
- CB100 and CD125S models ... 2.75 x 18 inch
- SL100 model ... 3.25 x 17 inch
- CB125S model ... 3.00 x 17 inch
- SL125 model ... 3.25 x 18 inch
- CD 125S model ... 2.75 x 18 inch

Tyre pressures
- Front ... 26 psi CB125S model
- Rear ... 28 psi CB125S model

1 General description

The wheel sizes on the CB, CD and CL range are all identical. The exception occurs in the SL range which have a larger diameter front wheel to give more ground clearance.

2 Front wheel - examination and removal

1 Place the machine on the centre stand so that the front wheel is raised clear of the ground. Spin the wheel and check for rim alignment. Small irregularities can be corrected by tightening the spokes in the affected area, although a certain amount of skill is necessary if over-correction is to be avoided. Any 'flats' in the wheel rim should be evident at the same time. These are more difficult to remove with any success and in most cases the wheel will have to be rebuilt on a new rim. Apart from the effect on stability, there is greater risk of damage to the tyre bead and walls if the machine is run with a deformed wheel, especially at high speeds.

2 Check for loose or broken spokes. Tapping the spokes is the best guide to the correctness of tension. A loose spoke will produce a quite different note and should be tightened by turning the nipple in an anticlockwise direction. Always check for run-out by spinning the wheel again.

3 If several spokes require retensioning or there is one that is particularly loose, it is advisable to remove the tyre and tube so that the end of each spoke that projects through the nipple after retensioning can be ground off. If this precaution is not taken,

2.4 This spacer must not be mislaid

3.6 Front brake shoes in location

3.7 Check surface for scoring or cracks

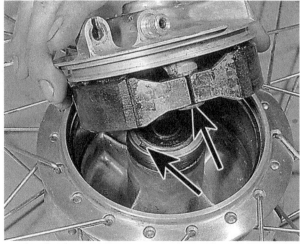

3.15 Relocate speedometer drive tabs with slots in hub

the portion of the spoke that projects may chafe the inner tube and cause a puncture.

4 Disconnect the speedometer drive by unscrewing the knurled nut and disconnect the front brake cable. Support the wheel, pull out the spindle and remove the wheel. Do not mislay the spacer from the left hand side.

3 Front brake assembly - examination, renovation and reassembly

1 Remove the front wheel and the brake plate as described in Section 2 of this Chapter.
2 Pull out the brake plate from the wheel.
3 Examine the condition of the brake linings. If they have worn thin or unevenly, the brake shoes should be renewed.
4 The brake shoes are the bonded type so renewal of the complete shoe is necessary if the lining is worn badly.
5 To remove the brake shoes from the brake plate, pull them apart whilst lifting them upward, in the form of a V. When they are clear of the brake plate, the return springs can be removed and the shoes separated.
6 Before replacing the brake shoes, check that the brake operating cam is working smoothly, and not binding in the pivot. The cam can be removed for cleaning and greasing by unscrewing the nut on the brake operating arm and drawing the arm off, after its position relative to the cam spindle has been marked so that it is replaced in exactly the same position. The spindle and cam can then be pressed out of the housing in the back of the brake plate.
7 Check the inner surface of the brake drum on which the brake shoes bear. The surface should be smooth and free from score marks or indentations, otherwise reduced braking efficiency is inevitable. Remove all traces of brake lining dust and wipe both the brake drum surface and the brake shoes with a clean rag soaked in petrol, to remove any traces of grease. Check that the brake shoes have chamfered ends to prevent pick-up or grab. Check that the brake shoe return springs are in good order and have not weakened.
8 If the brake drum is scored it is advisable to take it to a Honda dealer who may be able to remachine it. Not all dealers are able to do this but most may know of an engineering establishment which can skim the drum out. Sizes are given in the Specifications Section of this Chapter.
9 To reassemble the brake shoes on the brake plate, fit the return springs first and force the shoes apart, holding them in a V formation. If they are not located with the operating cam they can usually be snapped into position by pressing downward. Do not use excessive force or the shoes may distort permanently.
10 If the linings are found to be oily or greasy either the large

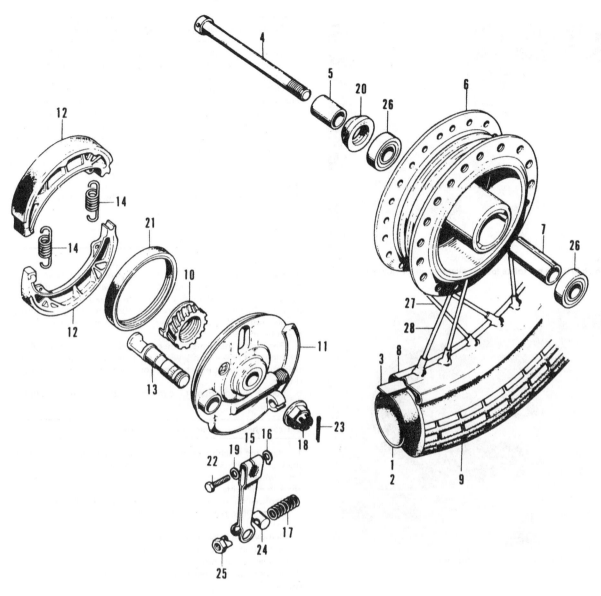

Fig. 5.1. Front wheel

1 Inner tube	8 Front wheel rim	15 Front brake operating arm	22 Hexagon bolt
2 Inner tube	9 Front wheel tyre	16 Locknut	23 Split pin
3 Rim tape	10 Speedometer drive gear	17 Brake arm return spring	24 Trunnion
4 Front wheel spindle	11 Front brake plate	18 Front wheel spindle nut	25 Brake adjusting nut
5 Spacer	12 Brake shoe - 2 off	19 Brake arm washer	26 Wheel bearing - 2 off
6 Front wheel hub	13 Front brake operating cam	20 Oil seal	27 Spoke - 18 off
7 Bearing spacer	14 Brake shoe spring - 2 off	21 Oil seal	28 Spoke - 18 off

oil seal stopping grease passing from the speedometer drive to the drum has failed or some inexperienced person has pumped far too much grease in the drive.

11 If the seal is weak, damaged or just worn, hook out the old and tap in the new one. Fit the correct way round, with the flat side of the seal facing outwards. Tap it in until it is flush with the edge.

12 Pull out the speedometer drive assembly. Check for wear and damage such as the drive pegs being bent back which locate the drive with the hub. These can be straightened to their original position so that they locate with the small slots of the hub.

13 If the gear is worn or has stripped renew it and check both the cable and speedometer head. This form of damage is usually caused by a stiff or seized component placing a heavy loading on the gear.

14 Regrease in moderation and replace.

15 When reassembling the wheel do not forget to ensure the locating tabs for the speedometer drive have seated before putting the wheel back into the machine, and that they have remained in position as the wheel is fitted.

Chapter 5/Wheels, brakes and tyres

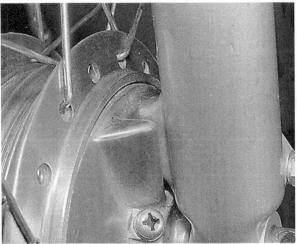

5.2 Notch in brake backplate locates over peg on fork lower leg

5.4 Replace the split pin after tightening

8.1 Rear brake shoes in location

4 Front wheel bearings - removal and replacement

1 There are two bearings in the front wheel. If the wheel has any side play when fitted to the machine or any roughness, the wheel bearings need to be renewed.
2 Using a small flat ended drift, place it inside the hub against one of the wheel bearings and tap the bearing out of position. Remove the central spacer and then tap out the other bearing from the other side. Drive the bearings outwards in each case.
3 Use a good quality grease and grease the new bearings. Tap in one bearing, turn the wheel over, insert the central spacer and then tap in the other bearing. Fit each bearing with the dust seal facing outwards and do not cut or damage it when tapping it home.
4 Replace the outer left hand oil seal which will have become dislodged. Renew it if necessary.

5 Front wheel - replacement

1 Locate the speedometer drive tabs with the slots in the hub.
2 Place the spacer on the left hand side of the wheel and then, holding the brake plate into the wheel place the wheel into the forks. The brake anchor slot and peg must locate correctly.
3 Hold the wheel in place and push the spindle through until it is right home. The spindle goes through the side opposite to the brake plate.
4 Reconnect the front brake cable and adjust the brake. Applying the front brake firmly with one hand, spin on the nut and tighten; when tight insert a new correct-sized split pin and bend it over.
5 Locate the speedometer cable and tighten the locking nut.

6 Rear wheel - examination and removal

1 Place the machine on the centre stand and before removing the rear wheel, follow the procedure described for checking the front wheel for alignment of the rim, loose or broken spokes or any other defects.
2 Remove the nut on the rear brake rod and disconnect it.
3 Using pliers, remove the rear chain link to disconnect the final drive chain, but do not pull off the gearbox sprocket.
4 Remove the brake plate torque arm fixing bolt.
5 Straighten out the split pin, pull it out and then unscrew the rear wheel spindle nut.
6 Support the rear wheel, pull out the spindle and remove the wheel.

7 Rear wheel bearings - removal and replacement

1 Remove the oil seal and remember it has to be replaced by a new one. If the bearings are worn or rough, both should be renewed.
2 Tap out the old bearings and fit new ones, using the method recommended for the front wheel bearings. Do not forget to pack with new, clean grease.

8 Rear brake - removal, examination and replacement

To check and replace the rear brake shoes refer to the Section that relates to the front brake shoes.

9 Rear wheel sprocket - examination and replacement

1 To remove the rear sprocket release with the aid of circlip pliers, the large circlip around the left hand side of the hub and straighten out the tab washers on the four nuts. Remove the nuts and then withdraw the rear wheel sprocket.

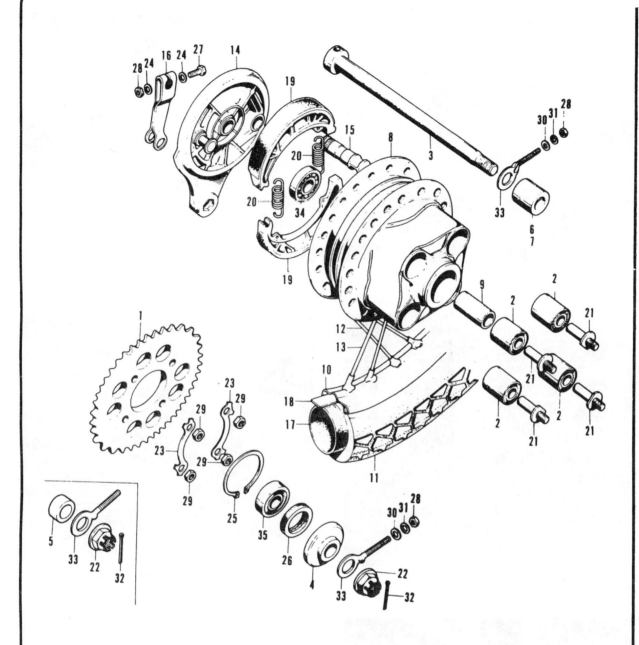

Fig. 5.2. Rear wheel

1 Rear wheel sprocket (40 teeth - standard size)
2 Cush drive rubbers - 4 off
3 Rear wheel spindle
4 Collar
5 Collar (alternative)
6 Brake plate collar
7 Brake plate collar
8 Rear wheel hub
9 Bearing spacer
10 Rear wheel rim
11 Rear wheel tyre
12 Spoke - 18 off
13 Spoke - 18 off
14 Rear brake plate
15 Rear brake operating cam
16 Rear brake operating arm
17 Inner tube
18 Rim tape
19 Brake shoe - 2 off
20 Brake shoe spring - 2 off
21 Sprocket retaining stud - 4 off
22 Rear wheel spindle nut
23 Tongued washer
24 Brake arm washer
25 Circlip
26 Oil seal
27 Hexagon bolt
28 Nut - 3 off
29 Nut - 4 off
30 Plain washer - 2 off
31 Spring washer - 2 off
32 Split pin
33 Chain adjuster - 2 off
34 Right-hand wheel bearing
35 Left-hand wheel bearing

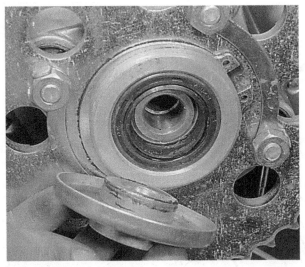
9.1 Inspect oil seal and bearing

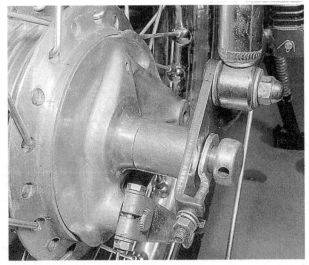
10.2 Rear wheel in correct location

10.4a Replace and adjust rear brake nut

10.4b Rear brake torque arm anchorage

11.6 Closed end must face direction of rotation

12.1 Line up marks equally each side and replace split pin

Chapter 5/Wheels, brakes and tyres

2 To examine the sprocket, clean it thoroughly and look for signs of wear such as hooked teeth, broken teeth or cracking. If, on checking, any of these defects are evident, a new replacement is necessary.
3 If the rear wheel sprocket is worn it is advisable to renew the gearbox sprocket and also the chain, otherwise the newly-fitted parts will have an accelerated rate of wear.
4 To replace the sprocket, place the four mounting studs, and using new tab washers, tighten up the nuts and bend the locking washers against them. Finally, replace the large circlip in the centre of the hub.
5 The cush drive bushes are a tight push fit in the hub. After a considerable length of time the bushes will corrode in position making removal almost impossible without the correct expanding extractor. It is recommended strongly that removal of the bushes is carried out only by a Honda specialist who will be able to accomplish the operation without risking damage to the wheel hub.

10 Rear wheel - replacement

1 To replace the rear wheel place the rear brake backplate into the brake drum, hold it in place and insert the assembly into the swinging arm of the machine.
2 Holding the wheel in place, take the spindle and feed it through a wheel adjuster (the correct way) then pass it through the swinging arm fork. Slip the spacer in between the brake plate and the frame, then finish pushing the spindle through. Place the other wheel adjuster over the end and then the wheel nut and washer.
3 Do not tighten yet but reconnect the brake plate torque arm and adjust the rear chain using the marks on the fork ends to aid correct alignment of the wheel.
4 Replace the rear brake nut on the brake rod and adjust. Apply the brake, tighten the wheel nut and replace the split pin. Tighten up the brake torque arm nut and bolt, then check all round again. Do not forget to replace the chain link in the correct position - with the closed end of the spring clip facing the direction of travel of the chain.

11 Final drive chain - examination and lubrication

1 The final drive chain is not fully enclosed. The only lubrication provided takes the form of periodic oiling during maintenance.
2 Chain adjustment is correct when there is approximately ¾ in play in the middle of the chain run, measured at either the top or the bottom of the run. Always check at the tightest spot of the chain with the rider seated normally.
3 To check whether the chain needs renewing, lay it lengthwise in a straight line and compress it endwise so that all play is taken up. Anchor one end firmly, then pull endwise in the opposite direction and measure the amlunt of stretch. If it exceeds ¼ inch per foot, renewal is necessary. Never use an old or worn chain when new sprockets are fitted; it is advisable to renew the chain at the same time so that all new parts run together.
4 Every 2000 miles remove the chain and clean it thoroughly in a bath of paraffin before immersing it in a special chain lubricant such as Linklyfe or Chainguard. These latter types of lubricant are applied in the molten state (the chain is immersed) and therefore achieve much better penetration of the chain links and rollers. Furthermore, the lubricant is less likely to be thrown off when the chain is in motion.
5 When replacing the chain, make sure that the spring link is positioned correctly, with the closed end facing the direction of travel. Replacement is made easier if the ends of the chain are pressed into the teeth of the rear wheel sprocket whilst the connecting link is inserted. Alternatively, a simple 'chain joiner' can be used.

12 Final drive chain - adjustment

1 Adjusting the drive chain is easy. Slacken the rear wheel nut (after removing the split pin) and turn the adjuster nuts half a turn at a time to move the wheel back until there is ¾ inch up and down movement in the chain's tightest spot with the rider seated on the machine.
2 Check that the wheel is in line, apply the rear brake and tighten the spindle nut; recheck to make sure that it has not moved. Replace the split pin and re-adjust the rear brake. The marks on the fork ends will show whether the wheel is aligned correctly.

13 Front and rear brake adjustment

1 Adjust the front brake by tightening the adjuster nut on the threaded portion of the front brake cable. Tighten it slowly and apply the front brake lever until the operating action seems correct. The lever must not touch the handlebars when the brake is applied fully nor must the brake bind when the lever is slack.
2 The rear brake is adjusted in much the same way, by screwing up the nut on the rod which passes through the rear brake arm until the brake has sufficient operating movement. Never adjust either brake until there is no movement at all; always check that the wheels spin freely after adjustment.

14 Front wheel balancing

1 The need may arise on some machines to balance the front wheel complete with tyre and tube. The out of balance forces which exist are then eliminated and the handling of the machine improved. A wheel which is badly out of balance produces throughout the steering, a most unpleasant hammering effect at high speeds.
2 One ounce and half ounce balance weights are available which can be slipped over the spokes and engaged with the square section of the spoke nipples. The balance weights are normally positioned diametrically opposite the tyre valve, which is usually responsible for the out of balance factor.
3 When the wheel is spun it will come to rest with the heaviest

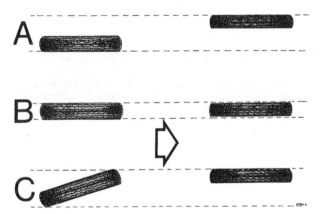

A & C Incorrect, B Correct

Fig. 5.3. Method of checking wheel alignment

point downward; balance weights should be added opposite to this point. Add or subtract balance weights until the wheel will rest in ANY position after it has been spun.
4 If balance weights are not available, wire solder wrapped around the spokes, close to the nipples, is an excellent substitute.
5 There is no necessity to balance the rear wheel for normal road use.

15 Tyres - removal and replacement

1 At some time or other the need will arise to remove and replace the tyres, either as a result of a puncture or because renewal is required to offset wear. To the inexperienced, tyre changing represents a formidable task yet if a few simple rules are observed and the technique learned the whole operation is surprisingly simple.
2 To remove the tyre from either wheel, first detach the wheel from the machine by following the procedure given in this Chapter whether the front or the rear wheel is involved. Deflate the tyre by removing the valve insert and when it is fully deflated push the bead of the tyre away from the wheel rim on both sides so that the bead enters the centre well of the rim. Remove the locking cap and push the tyre valve into the tyre.
3 Insert a tyre lever close to the valve and lever the edge of the tyre over the outside of the wheel rim. Very little force should be necessary; if resistance is encountered it is probably due to the fact that the tyre beads have not entered the well of the wheel rim all the way round the tyre.
4 Once the tyre has been edged over the wheel rim, it is easy to work around the wheel rim so that the tyre is completely free on one side. At this stage, the inner tube can be removed.
5 Working from the other side of the wheel, ease the other edge of the tyre over the outside of the wheel rim furthest away. Continue to work around the rim until the tyre is free completely from the rim.
6 If a puncture has necessitated the removal of the tyre, re-inflate the inner tube and immerse it in a bowl of water to trace the source of the leak. Mark its position and deflate the tube. Dry the tube and clean the area around the puncture with a petrol soaked rag. When the surface has dried, apply rubber solution and allow this to dry before removing the backing from a patch and applying the patch to the surface.
7 It is best to use a patch of the self-vulcanising type, which will form a very permanent repair. Note that it may be necessary to remove a protective covering from the top surface of the patch, after it has sealed in position. Inner tubes made from synthetic rubber may require a special type of patch and adhesive if a satisfactory bond is to be achieved.
8 Before replacing the tyre, check the inside of it to make sure that the agent which caused the puncture is not trapped. Check the outside of the tyre, particularly the tread area, to make sure nothing is trapped that may cause a further puncture.
9 If the inner tube has been patched on a number of past occasions, or if there is a tear or large hole, it is preferable to discard it and fit a new tube. Sudden deflation may cause an accident, particularly if it occurs in the front wheel.
10 To replace the tyre, inflate the inner tube just sufficiently for it to assume a circular shape. Then push it into the tyre so that it is enclosed completely. Lay the tyre on the wheel at an angle and insert the valve through the rim tape and the hole in the wheel rim. Attach the locking cap on the first few threads, sufficient to hold the valve captive in its correct location.
11 Starting at the point furthest away from the valve, push the tyre bead over the edge of the wheel rim until it is located in the central well. Continue to work around the tyre in this fashion until the whole of one side of the tyre is on the rim. It may be necessary to use a tyre lever during the final stages.
12 Make sure that there is no pull on the tyre valve and again commencing with the area furthest from the valve, ease the other bead of the tyre over the edge of the rim. Finish with the area close to the valve, pushing the valve up into the tyre until the locking cap touches the rim. This will ensure the inner tube is not trapped when the last section of the bead is edged over the rim with a tyre lever.
13 Check that the inner tube is not trapped at any point. Re-inflate the inner tube, and check that the tyre is seating correctly around the wall of the tyre on both sides, which sould be equidistant from the wheel rim at all points. If the tyre is unevenly located on the rim, try bouncing the wheel when the tyre is at the recommended pressure. It is probable that one of the beads has not pulled clear of the centre well.
14 Always run the tyres at the recommended pressures and never under or over-inflate. See Specifications for recommended pressures.
15 Tyre replacement is aided by dusting the side walls, particularly in the vicinity of the beads, with a liberal coating of French chalk. Washing up liquid can also be used to good effect, but this has the disadvantage of causing the inner surfaces of the wheel rim to rust.
16 Never replace the inner tube and tyre without the rim tape in position. If this precaution is overlooked there is good chance of the ends of the spoke nipples chafing the inner tube and causing a crop of punctures.
17 Never fit a tyre which has a damaged tread or side walls. Apart from the legal aspects there is a very great risk of a blow-out, which can have serious consequences on any two wheel vehicle.
18 Tyre valves rarely give trouble but it is always advisable to check whether the valve itself is leaking before removing the tyre. Do not forget to fit the dust cap which forms an effective second seal. This is especially important on a high performance machine, where centrifugal force can cause the valve insert to retract and the tyre to deflate without warning.

16 Security bolts

1 If the drive from a high powered engine is applied suddenly to the rear wheel or if the machine has an especially low bottom gear ratio, wheel spin will occur on a loose or slippery surface with an initial tendency for the wheel rim to creep in relation to the tyre and inner tube. Under these circumstances there is risk of the valve being torn from the inner tube, causing the tyre to deflate rapidly, unless movement between the rim and tyre can be restrained in some way. A security bolt fulfills this role in a simple and effective manner, by clamping the bead of the tyre to the well of the wheel rim so that any such movement is no longer possible.
2 Although the wheel rims of the Honda SL trail models are ridged on their inside edges to achieve much the same effect, it is advantageous to fit a security bolt to the rear wheel if the machine is used for serious competition or off-road use. The accompanying illustrations show how the security bolt (supplied as an optional extra) is fitted and secured. Note that before attempting to remove or replace a tyre, the security bolt must be slackened off so that the clamping action is released.

Fault diagnosis is on page 78

Tyre changing sequence - tubed tyres

 Deflate tyre. After pushing tyre beads away from rim flanges push tyre bead into well of rim at point opposite valve. Insert tyre lever adjacent to valve and work bead over edge of rim.

 Use two levers to work bead over edge of rim. Note use of rim protectors

 Remove inner tube from tyre

 When first bead is clear, remove tyre as shown

 When fitting, partially inflate inner tube and insert in tyre

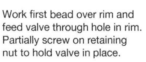

 Work first bead over rim and feed valve through hole in rim. Partially screw on retaining nut to hold valve in place.

 Check that inner tube is positioned correctly and work second bead over rim using tyre levers. Start at a point opposite valve.

Work final area of bead over rim whilst pushing valve inwards to ensure that inner tube is not trapped

Chapter 5/Wheels, brakes and tyres

17 Fault diagnosis - wheels, brakes and tyres

Symptom	Cause	Remedy
Ineffective brake	Worn brake linings	Replace.
	Foreign bodies on brake lining surface	Clean.
	Incorrect engagement of brake arm serration	Reset correctly.
	Worn brake cam	Replace.
Handlebars oscillate at low speeds	Buckle or flat in wheel rim, most likely front wheel	Check rim alignment by spinning wheel. Correct by retensioning spokes or building on new rim.
	Tyre not straight on rim	Check tyre alignment.
Machine lacks power and poor acceleration	Brakes binding	Warm brake drum provides best evidence. Re-adjust brakes.
Brakes grab when applied gently	Ends of brake shoes not chamfered	Chamfer with file.
	Elliptical brake drum	Lightly skim on lathe.
Brake pull-off spongy	Brake cam binding in housing	Free and grease.
	Weak brake shoe springs	Renew if springs have not become displaced.
Harsh transmission	Worn or badly adjusted final drive chain	Adjust or renew.
	Hooked or badly worn sprockets	Renew as a pair.
	Loose rear sprocket	Check bolts.

Chapter 6 Electrical system

Contents

General description ... 1	Flashing indicator lamps - replacing bulbs ... 11
Crankshaft alternator - checking the output ... 2	Flasher unit - location and replacement ... 12
Battery - charging procedure and maintenance ... 3	Headlamp dipswitch and pilot lamp ... 13
Rectifier (Selenium type) - general ... 4	Horn push and horn - adjustment ... 14
Resistor - function and location ... 5	Fuse - location and replacement ... 15
Headlamp - replacing bulbs and adjusting beam height ... 6	Ignition switch ... 16
Tail and stop lamp - replacing bulbs ... 7	Emergency switch or ignition cut-out ... 17
Rear brake stop lamp switch - adjusting ... 8	Headlamp switch ... 18
Front brake stop lamp switch - location and replacement ... 9	Wiring - layout and examination ... 19
Speedometer and tachometer indicator lamps - replacing bulbs 10	Fault diagnosis - electrical system ... 20

Specifications

Battery
Voltage	6 volts
Make	Yuasa
Type	6N6-3B
Capacity	6 amp hr
Earth connection	Negative

Fuse
Rating:
CB125S and CD125S models	10 amps
CB100, CL100, SL100 and SL125 models	15 amps

Bulbs
Main headlamp:
CB100, CL100, SL100 and SL125 models	35/25W
CB125S and CD125S models	25/25W

Tail/stop lamp:
CB100, CL100, SL100 and SL125 models	5.3/17W
CB125S and CD125S models	3/10W

Flashing indicators:
CB100 and CL100 models	18W
CB125S and CD125S models	18W

Speedometer (and tachometer) indicator lamps:
All models	1.5W

1 General description

The electrical system is 6 volt and comprises a battery, which in turn receives a charge from an alternator mounted on the left hand end of the crankshaft via a rectifier. The alternator takes the form of a rotary magnet revolving at engine speed within six equispaced coils which do not actually touch the magnet. Output is controlled by a resistor, which is switched into the circuit to dissipate excess current in the form of heat.

2 Crankshaft alternator - checking the output

1 An ammeter and a voltmeter of the centre zero type are required for this test, both of the moving coil type. The ammeter should have a 0 - 10 volt range and the ammeter 0 - 2 amps. Before the check is carried out, make sure that the battery is fully charged (specific gravity of the electrolyte 1.28 at 20°C).
2 Disconnect the red/white lead from the positive terminal of the battery and connect it to the negative terminal of the

		Lighting switch	Dimmer switch	Initial charging rpm		5000 rpm	
				rpm	Battery voltage	Charging current	Battery voltage
100 cc Series	Day	OFF	OFF	1000 rpm	6.8V	1.3A	7.8V
	Night	ON	HB (high beam)	3500 rpm	6.8V	1.3A	7.8V
		ON	LB (low beam)	2200 rpm	6.8V	1.3A	7.2V
125 cc Series	Day	OFF	OFF	1000 rpm	6.8V	1.7A	7.9V
	Night	ON	LB (low beam)	2000 rpm	6.8V	1.3A	7.8V

ammeter. Connect the positive terminal of the battery to the negative terminal of the ammeter. Take a further connection from the red/white lead to the positive terminal of the voltmeter and connect the negative terminal of the voltmeter to a good earthing point on the frame of the machine. Start the engine.

3 Measure the battery voltage and the charging current. They should correspond with the figures given in the table above. If they are less than the values shown, further advice should be sought from either a Honda dealer or an auto-electrical expert.

3 Battery - charging procedure and maintenance

1 Whilst the machine is used on the road it is unlikely that the battery will require attention other than routine maintenance because the generator will keep it fully charged. However, if the machine is used for a succession of short journeys only, mainly during the hours of darkness when the lights are in full use, it is possible that the output from the generator may fail to keep pace with the heavy electrical demand, especially if the machine is parked with the lights switched on. Under these circumstances it will be necessary to remove the battery from time to time to have it charged independently.

2 The battery is located behind the left panel in a carrier slung on the left hand side of the machine, under the side cover. It is secured by a strap which, when released will permit the battery to be withdrawn after disconnection of the leads. The battery negative is always earthed.

3 The normal off-machine charge rate is 0.6 amp. A more rapid charge up to 2 amps can be given in an emergency, but this should be avoided if possible because it will shorten the life of the battery. See accompanying table.

4 When the battery is removed from the machine, clean the battery top. If the terminals are corroded scrape them clean and cover them with Vaseline (not grease) to protect them from further attack. If a vent tube is fitted, make sure it is not obstructed and that it is arranged so that it will not discharge over any parts of the machine.

5 If the machine is laid up for any period of time, the battery should be removed and given a 'refresher' charge every six weeks or so, in order to maintain it in good condition.

4 Rectifier (Selenium type) - general

1 The rectifier is bolted on the left hand side of the machine, to the right of the battery; its function is to convert the alternating current from the alternator to the direct current which can then be used to charge the battery and operate the electrical system.

2 The rectifier is deliberately placed in this location so that it is not exposed directly to water or oil and yet has free circulation of air to permit cooling. It should be kept clean and dry; the nuts connecting the rectifier plates should not be disturbed under any circumstances.

3 It is not possible to check whether the rectifier is functioning correctly without the appropriate test equipment. If performance is suspect a Honda agent or auto-electrical expert should be consulted. Note that the rectifier will be destroyed if it is subjected to a reverse flow of current.

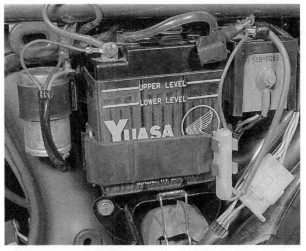

3.2. The electrical compartment showing battery, flasher unit, rectifier and fuse

4.2 The rectifier location

Chapter 6/Electrical system

Fig. 6.1. Battery charging instructions

	Normal charge	Rapid charge
Charging current rate	0.6AH	2.0AH max.
Checking for full charge	(1) Specific gravity: 1.260 - 1.280 (20º C: 68º F) maintained constant (2) 0.2AH – 0.6AH (3) 7.5V – 8.3V	(1) Specific gravity: 1.260 - 1.280 maintained at 20º C (68º F) (2) Voltage: When large volume of gas is emitted from the battery (in about 2-3 hours for fully discharged battery), reduce charging rate to 0.2A. Battery is fully charged when a voltage of 7.5V is maintained.
Charging duration	By this method, a battery with specific gravity of electrolyte below 1.220 at 20º C (68º F) will be fully charged in approximately 12-13 hours.	By this method, battery with specific gravity of electrolyte below 1.220 at 20º C (68º F) will be fully charged approximately 1-2 hours.
Remarks		When the charging is urgent, quick charging method may be used, however the recommended charging current rate should be under 2.0A.

Note: Battery should not be charged near an open fire
Terminals should be cleaned with clean water unless corroded. Apply petroleum jelly afterwards.

4 When tightening the rectifier securing nut, hold the nut at the other end with a spanner. Apart from the fact that the securing stud is sheared very easily if overtightened, there is risk of the plates twisting and severing their internal connections.

5 Resistor - function and location

1 The resistor is situated in a protective cage bolted under the bottom yoke of the front forks. The resistor's function is to drain off excess power generated when only the side lights are in use, and to convert it to heat which is dissipated by the air stream. When the main headlamp is in use, the resistor is switched out of the circuit so that full current is available to meet the increased electrical demand.

2 If the pilot light burns out with recurring frequency but the headlamp is unaffected, the resistor should be suspected. As it is a sealed unit a new replacement should be fitted.

6 Headlamp - replacing bulbs and adjusting beam height

1 To replace the headlamp bulb, slacken the rim retaining screws in the lower side of the headlamp shell and using both hands to clasp the unit, pull the bottom out first. The unit will then lift away.

2 Unplug the bulb holder from the rear of the reflector unit and take out the bulb. Use it as reference to get the correct voltage and wattage if a replacement is needed as there are numerous types and fittings available.

3 The bulb can only be replaced one way because of the design of the base. Replace the bulb and the holder, then clip the top

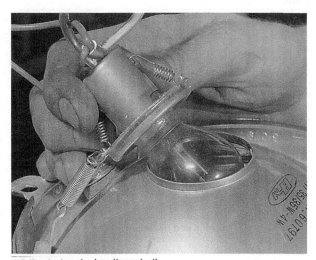

6.2 Replacing the headlamp bulb

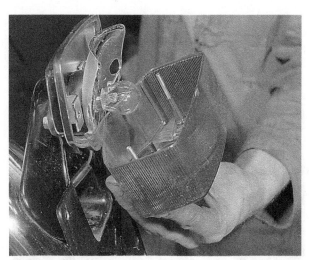

7.1 Rear bulb replacement

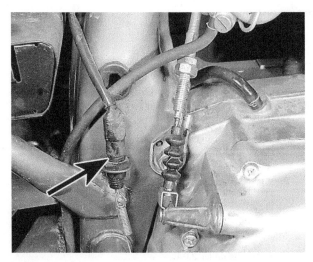

8.1 Brake stop lamp switch location

11.2 Lens is retained by two screws

12.1 The flasher unit location

part of the rim into the headlamp and press in the bottom. Line up the holes and replace and tighten the fixing screws.
4 To adjust the beam height, sit on the machine with the lights on and the machine off the centre stand, and then switch to main beam. If the beam is shining higher than the headlamp itself, slacken the bolts which hold the shell and tilt the whole unit downwards until the beam shines just below headlamp height. Retighten the fixing bolts, taking care not to move the shell and recheck the setting.

7 Tail and stop lamp - replacing bulbs

1 The combined tail and stop lamp is fitted with a double filament bulb having offset pins to prevent its unintentional reversal in the bulb holder. The lamp unit serves a two-fold purpose - to illuminate the rear of the machine and the rear number plate, and to give visual warning when the rear brake is applied.
2 To gain access to the bulb, remove the two screws holding the red plastic lens in position. The bulb is released by pressing inwards and with an anticlockwise turning action, the bulb will now come out.

8 Rear brake stop lamp switch - adjusting

To adjust the stop lamp switch hold the switch body and with the correct size spanner screw down the adjuster nut thus raising the rear brake light switch and making it operate sooner. Never adjust too tightly, the light should come on just as the first braking pressure is felt.

9 Front brake stop lamp switch - location and replacement

1 In order to give as much warning as possible to other road users of the braking of the machine, a front brake switch is incorporated in the front brake lever assembly.
2 There is no means of adjustment if it malfunctions, in which case the switch must be renewed.
3 To replace the switch, the headlamp reflector unit must be taken off and the wire from the switch disconnected. Take off the front brake lever and feed the switch and wires back out towards the lever pivot. Replace the new switch in the reverse order.

10 Speedometer and tachometer indicator lamps - replacing bulbs

Pull out the rubber bulb holder and twist the bulb to remove it from the bulb holder. This applies to all the bulbs in the instrument heads.

11 Flashing indicator lamps - replacing bulbs

1 All models (except the SL trail bikes) are fitted with flashing indicators, two attached to the front end of the machine and two to the rear. They are operated by a thumb switch on the handlebars.
2 To replace the bulbs in either back or front flasher, remove the two screws that retain the orange plastic lens then push and twist to remove the old bulb. Insert the new replacement. Be careful not to overtighten the lens retaining screws when refitting.

12 Flasher unit - location and replacement

1 The flasher unit is located on the left hand side of the machine behind the battery cover, to the front of the battery itself. It is circular in profile and is secured by a rubber band clip.
2 If malfunctioning occurs, the flasher unit must be renewed

Chapter 6/Electrical system

since it is a sealed unit and repairs are impracticable. It is easily removed by unplugging the wires and releasing the retaining clip. Before condemning the unit, make sure a defective bulb or connection is not the cause of the trouble.

3 All the wires are colour-coded so that reconnection of the new replacement causes no problems.

13 Headlamp dip switch and pilot lamp

1 The headlamp dip switch is incorporated in the handlebar controls and is used to dip the main headlamp beam to avoid dazzling oncoming traffic.

2 Should this unit malfunction it is necessary to fit a new replacement since repairs are usually impracticable. A sudden failure when changing from one beam to the other can plunge the lighting system into darkness, with disastrous consequences.

3 This switch also operates the pilot light as well. As long as the main switch is in the on position there is always a light showing.

14 Horn push and horn - adjustment

1 The horn push button is incorporated in the handlebar control and must be replaced if it malfunctions.

2 The horn is situated below the petrol tank at the front and has an adjustment screw inset into the rear. This is a volume screw which may need adjustment from time to time to compensate for wear inside the horn.

3 To adjust the horn, depress the horn button (with the ignition on) and screw the adjuster in or out to obtain maximum horn volume. Turn off the ignition when adjustment is correct.

15 Fuse - location and replacement

1 The fuse is incorporated in the red lead wire from the battery, enclosed in a nylon case. It is incorporated to protect the wiring and electrical components from accidental damage should a short circuit occur.

2 If the electrical system will not operate, a blown fuse should be suspected, but before the fuse is replaced the electrical system should be inspected to trace the reason for the failure of the fuse. If this precaution is not observed the replacement will almost certainly blow too.

3 The fuse is rated at 10 or 15 amps (see Specifications) and at least one spare should be carried at all times. In an extreme emergency and only when the cause of the failure has been rectified and no spare is available, a get-you-home repair can be made by wrapping silver paper around the blown fuse and re-inserting it in the fuse holder. It must be stressed that this is only an emergency measure and the fuse should be replaced at the earliest possible opportunity, as it affords no protection at all to the electrical system when bridged in this fashion.

16 Ignition switch

1 The ignition switch is fitted below the lower left hand side of the petrol tank close to the cylinder head. The key, when inserted, will complete the ignition and electrical circuits as the switch is turned.

2 Should the switch in any way malfunction a replacement should be obtained. This will necessitate a change of key, which is normally supplied with the new switch unit.

17 Emergency switch or ignition cut-out

1 The emergency switch is fitted to the right handlebar and is used only in emergencies. It completely breaks the ignition circuit should the throttle jam open or any other untoward thing happen. Should it malfunction, renewal is necessary.

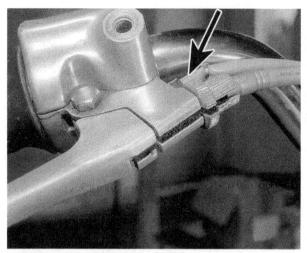

9.1 Front brake switch location

5.1 The resistor is mounted below the lower fork yoke

2 Care should be taken that the emergency switch is not turned to the 'off' position by accident, as this will immobilise the machine very effectively.

18 Headlamp switch

The headlamp switch is fitted into the right hand handlebar control and must be renewed if it malfunctions. As with the other switch gear, it is seldom practicable to effect a permanent repair.

19 Wiring - layout and examination

1 The cables of the wiring harness are colour-coded and will correspond with the accompanying wiring diagrams.

2 Visual inspection will show whether any breaks or frayed outer coverings are giving rise to short circuits which will cause the main fuse to blow. Another source of trouble is the snap connectors and spade terminals, which may make a poor connection if they are not pushed home fully.

3 Intermittent short circuits can sometimes be traced to a chafed wire passing through, or close to, a metal component, such as a frame member. Avoid tight bends in the cables or situations where the cables can be trapped or stretched, especially in the vicinity of the handlebars or steering head.

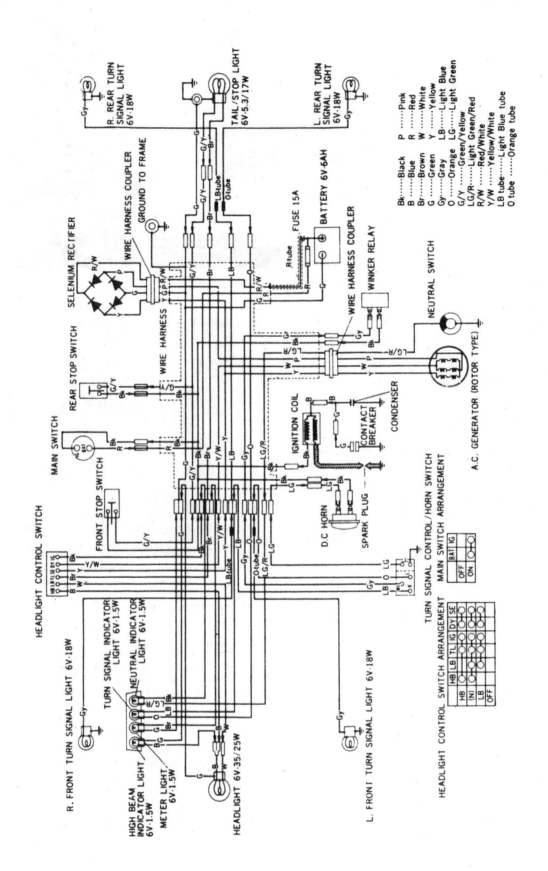

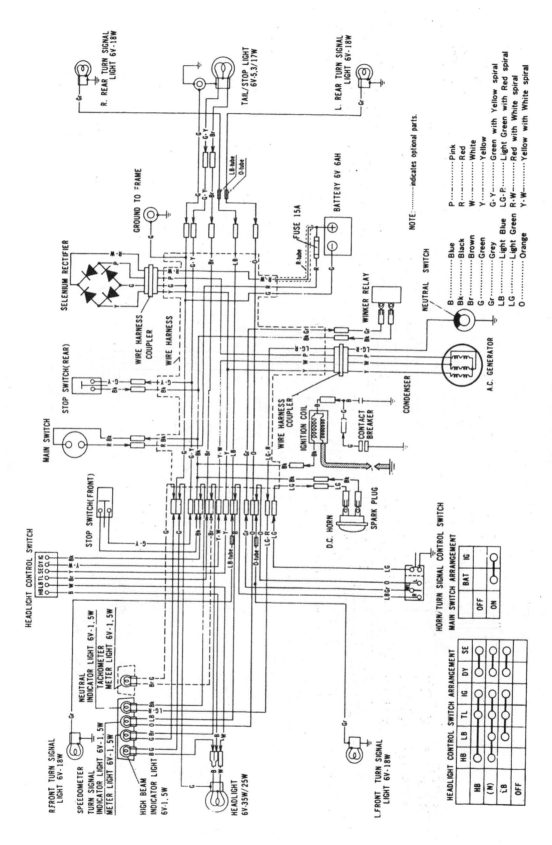

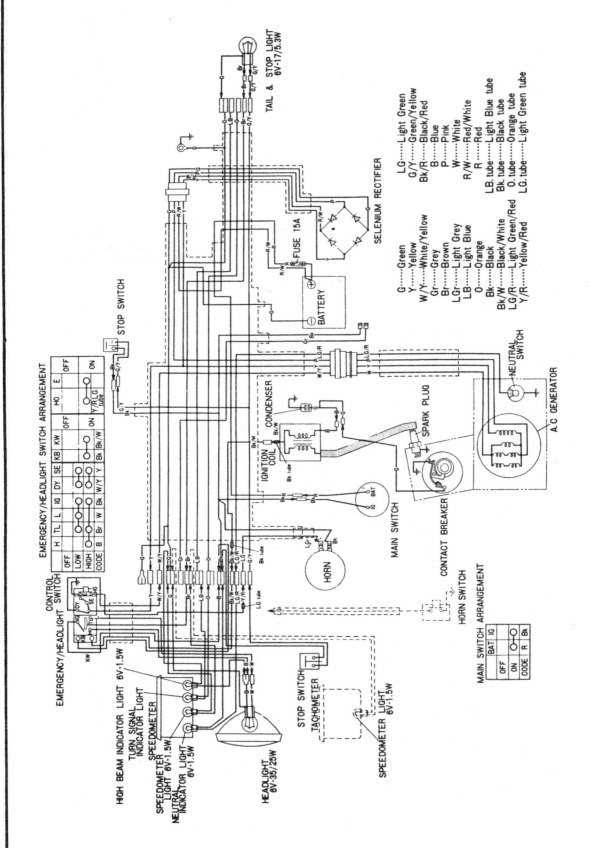

Wiring diagram - SL100

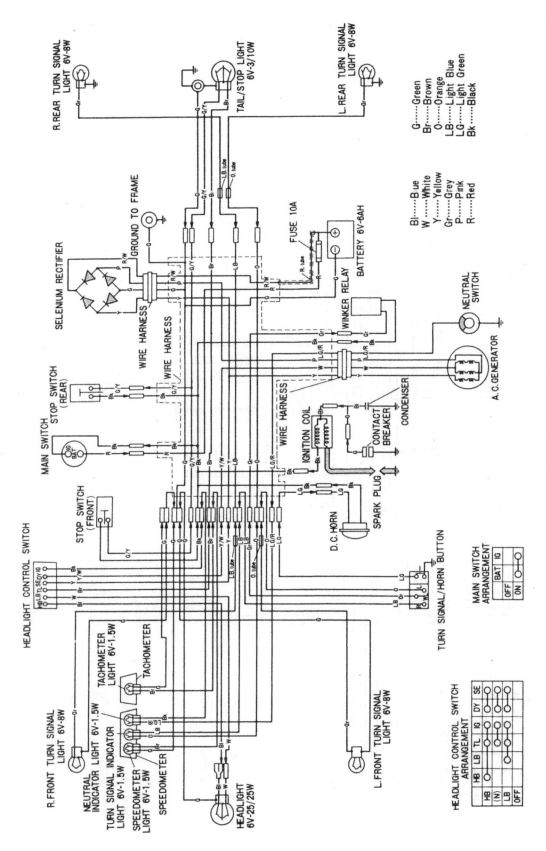

Wiring diagram - CB125S

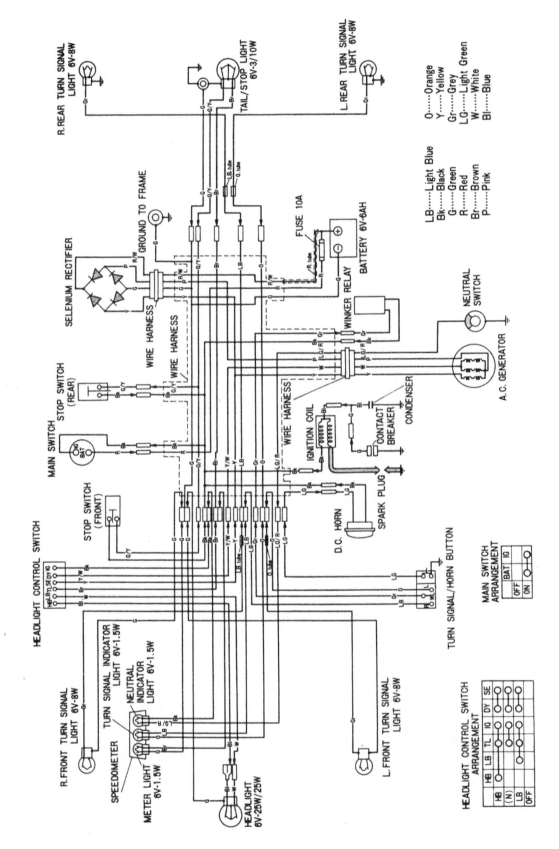

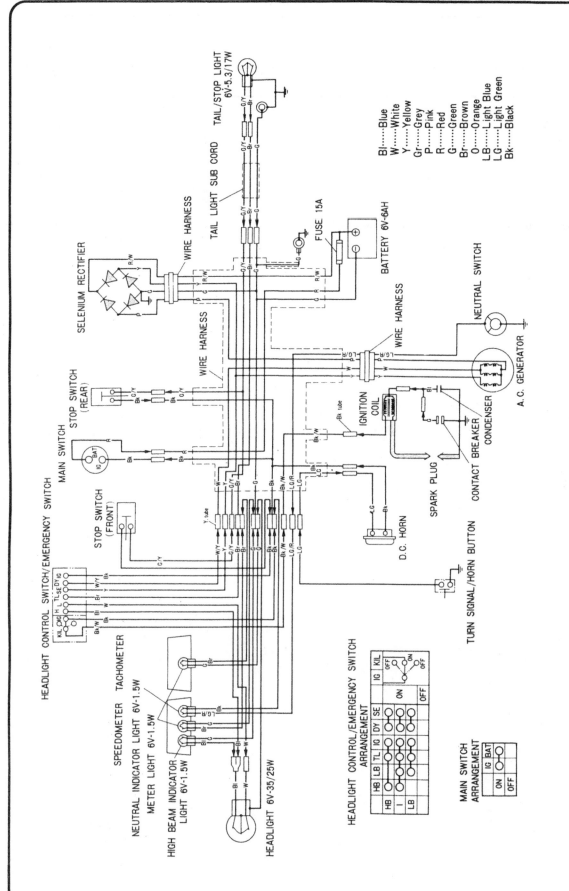

Wiring diagram - SL125

20 Fault diagnosis - electrical system

Symptom	Cause	Remedy
Complete electrical failure	Blown fuse	Check wiring and electrical components for short circuit before fitting new 15 amp fuse. Check battery connections, also whether connections show signs of corrosion.
	Isolated battery	
Dim lights, horn inoperative	Discharged battery	Recharge battery with battery charger and check whether alternator is giving correct output.
Constantly 'blowing' bulbs	Vibration, poor earth connection	Check whether bulb holders are secured correctly. Check earth return or connections to frame. Check resistor if pilot lamp bulb fails regularly.

Chapter 7 The CB125J model (CB125S-76)

Contents

Engine unit and cycle parts: dismantling, examination and reassembly ... 1
Dismantling the cylinder head ... 2
Crankshaft assembly ... 3
Right-hand side cover (crankcase): clutch operation ... 4
Petrol tank ... 5
Air cleaner assembly: dismantling and cleaning ... 6
Exhaust system ... 7
Front forks: headlamp mounting and internal construction ... 8
Rear suspension units: dismantling ... 9
Speedometer drive gearbox: location and examination ... 10
Front wheel disc brake assembly: examination, dismantling, renovation and reassembly ... 11
Front brake disc: removal, examination and replacement ... 12
Wheel bearings: examination and replacement ... 13
Front wheel disc brake: adjustment ... 14
Other minor differences - CB125J model ... 15

The data reproduced below relates only to the CB125J model (CB125S-76) where it differs from the information given in earlier Chapters. If the desired information is not available in these specifications, it may be assumed that the original data relating to the CB125S model, still applies.

Model dimensions
Overall length ... 1860 mm (73.2 in)
Overall width ... 750 mm (29.5 in)
Overall height ... 1050 mm (41.3 in)
Wheelbase ... 1205 mm (47.4 in)
Weight ... 93 kg (205 lb)

Quick reference maintenance data
Engine oil capacity ... 1 litre/1.1 US quart/0.9 Imp. quart
Fuel tank capacity ... 9.5 litres/2.5 US gallons/2.1 Imp. gallons
Fuel reserve ... 2.2 litres/0.6 US gallon/0.5 Imp. gallon
Spark plug ... NGK D8ES-L or Nippon Denso X24ES
Tyre pressures ... 26 psi front, 32 psi rear

Specifications
Compression ratio ... 9.4 : 1
Capacity ... 124 cc (7.6 cu in)

Gear ratios:
  Primary reduction ... 4.055
  Final reduction ... 2.333
  1st ... 2.769
  2nd ... 1.882
  3rd ... 1.450
  4th ... 1.173
  5th ... 1.000

Drive sprocket ... 15 teeth
Driven sprocket ... 35 teeth

Cylinder bore internal diameter
  Standard value ... 56.50 - 56.51 mm (2.2244 - 2.2247 in)
  Wear limit ... 56.60 mm (2.2283 in)

Piston outside diameter
  Standard value ... 56.46 - 56.48 mm (2.2228 - 2.2236 in)
  Wear limit ... 56.35 mm (2.2184 in)

Chapter 7: The CB125J model (CB125S-76)

Crankshaft runout
- Standard value, L and R ... 0.02 mm (0.00079 in) max
- Wear limit ... 0.05 mm (0.00179 in)

Connecting rod side clearance
- Standard value ... 0.05 - 0.35 mm (0.00197 - 0.01378 in) max
- Wear limit ... 0.8 mm (0.03150 in)

Bearing clearance
- Standard value ... 0 - 0.08 mm (0 - 0.000315 in) max
- Wear limit ... 0.05 mm (0.00179 in)

Valve stem outside diameter
- Inlet
  - Standard value ... 5.450 - 5.465 mm (0.2146 - 0.2152 in)
  - Wear limit ... 5.42 mm (0.2134 in)
- Exhaust
  - Standard value ... 5.430 - 5.445 mm (0.2138 - 0.2144 in)
  - Wear limit ... 5.4 mm (0.2126 in)

Valve guide - internal diameter
- Standard value ... 5.475 - 5.485 mm (0.2155 - 0.2159 in)
- Wear limit ... 5.5 mm (0.2165 in)

Valve spring - free length
- Outer
  - Standard value ... 44.85 mm (1.7658 in)
  - Wear limit ... 40.5 mm (1.5945 in)
- Inner
  - Standard value ... 39.2 mm (1.5433 in)
  - Wear limit ... 35.2 mm (1.3858 in)

Fork spring length
- Standard value ... 411.6 mm (16.2046 in)
- Wear limit ... 390 mm (15.3543 in)

Fork oil content per leg ... 105 - 110 cc ATF

Fork damper piston outside diameter
- Standard value ... 30.936 - 30.975 mm (1.21795 - 1.2195 in)
- Wear limit ... 30.9 mm (1.2165 in)

Carburettor settings:
- Main jet ... 110
- Slow running jet ... 45
- Needle position ... 2nd notch
- Pilot screw ... 1½ turns out
- Float level ... 24 mm
- Idle speed ... 1,300 rpm

Tyre sizes
- Front ... 2.75 x 18
- Rear ... 3.00 x 17

Alternator output ... 0.076 kw/5,000 rpm

In the UK a certain amount of confusion has occurred with regard to the coding of this model, the more so because it is officially listed as the CB125S-76 model in most other countries, including the USA. The transfers applied to the right-hand side cover conform to this latter code. However, there are certain differences between this model and the earlier CB125S versions which, no doubt, necessitated the use of the different code in the UK, for identification purposes. These include a cable-operated front wheel disc brake, a redesigned cylinder head that has a separate cambox cover casting and a different form of headlamp mounting, which permits the use of 'slimline' forks, without shrouds. A certain amount of restyling accounts for a redesigned fuel tank, with a recessed filler covered by a lockable flap, and, on some versions, a dualseat covered in a brown coloured material, as well as other superficial differences in appearance.

1 Engine unit and cycle parts: dismantling, examination and reassembly

1 Although the CB125J model is similar in a great many respects to the earlier CB125S model described in the preceding Chapters, reference should always be made to this Chapter first, in view of the need to follow a modified procedure when certain components of the CB125J model have to be removed and replaced. Where no information is given in this Chapter, it may be assumed that the procedure is identical to that given for the CB125S model in the earlier Chapters.

2 Routine Maintenance procedure closely follows the schedules already described in the earlier part of this Manual.

The CB 125J model (CB 125S - 76)

Chapter 7: The CB125J model (CB125S-76)

2 Dismantling the cylinder head

1 Although the complete engine unit can be lifted out of the frame with the cylinder head in situ, and the cylinder head itself subsequently removed as a complete unit, after the camshaft chain and sprocket have been detached, it is necessary to remove the cambox cover for access to the valve gear.

2 The cambox cover is retained to the cylinder head by four short socket head screws. When these are removed and the cylinder head nuts, the cover can be lifted off, to expose the valve gear. Note that there are two dowels to locate the cover correctly and no gasket is required at the jointing surface. It is assumed, of course, that the cylinder head will have been removed from the cylinder barrel first by unscrewing the four domed nuts and the single long bolt that retains it to the main engine assembly.

3 When the cambox cover has been removed, the camshaft can be lifted out from its bearings and the rocker arm spindles withdrawn, to release the rocker arms. The rocker arm spindles are retained by a locking plate secured by a single screw. Each spindle is provided with an internal thread, so that a bolt can be screwed in and used to pull them out. Note the recess milled in each spindle, which must line up with the hole through

2.2 The cambox cover is separate, retained by socket head screws and domed nuts

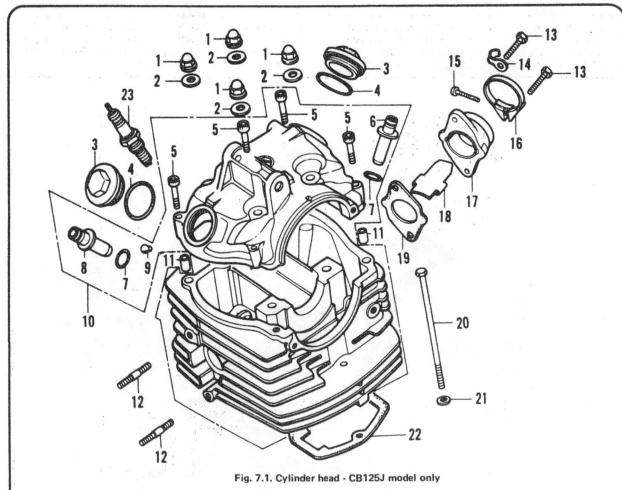

Fig. 7.1. Cylinder head - CB125J model only

1   Dome nut - 4 off
2   Sealing washer - 4 off
3   Valve clearance adjusting cap - 2 off
4   'O' ring - 2 off
5   Socket screw - 4 off
6   Inlet valve guide
7   'O' ring - 2 off
8   Exhaust valve guide
9   Rubber oil seal
10 Cylinder head assembly complete
11 Dowel - 2 off
12 Exhaust port stud - 2 off
13 Bolt - 2 off
14 HT lead clip
15 Hose clip screw
16 Hose clip
17 Induction stub
18 Atomiser
19 Heat insulator
20 Bolt
21 Plain washer
22 Cylinder head gasket
23 Spark plug

Chapter 7: The CB125J model (CB125S-76)

which the cylinder holding down studs pass, when they are replaced.

4 There is no necessity to remove the rocker spindles and arms unless these parts are to be examined for wear or if it is necessary to remove the valves. Removal of the rocker arms will give unimpeded access to the valve springs and make the whole dismantling and reassembly operation very much easier.

5 For valve clearance adjustment, the cambox cover can be left in situ. Access to the valve adjusters is available through threaded circular covers in the casting, one close to each valve. The clearance can be adjusted and set whilst the cylinder head is off the machine, since the camshaft itself is contained within the same casting.

6 Modification of the cam chain tensioner components has resulted in a change of tension adjustment procedure. To adjust the tension, start the engine and allow it to run at a fast tick-over. Prise the rubber boot from the tensioner push-rod housing and slacken the housing by applying a spanner to the large hexagon. The tension will adjust automatically. Retighten the housing. This will cause the two angled bushes to lock the tensioner pushrod in position, so maintaining the pre-set tension until adjustment is required again.

3 Crankshaft assembly

1 The crankshaft assembly is virtually identical with that of the earlier CB125S model. There is, however, no thrust washer behind the right-hand main bearing and both main bearings are of the radial ball type.

4 Right-hand side cover: clutch operation

1 A redesigned right-hand side cover permits the use of a different design of clutch operating arm, even though the clutch assembly is virtually identical with that of the CB125S model. The arm passes through the top of the cover in the vertical plane and is spring-loaded, with a built-in stop. A small detachable cable stop is located close to the right-hand corner of the crankcase/cylinder barrel flange joint and secured by means of the upper front crankcase bolt, which passes through its centre. The inner cable extends from this stop to the end of the operating arm, which is at right angles to and an integral part of, the clutch operating shaft.

2 This revised clutch operating arrangement obviates the need for the clutch adjustment screw, found in the lower rear portion of the cover of the earlier CB125S models.

5 Petrol tank

1 Although of different shape from that fitted to the CB125S model, the fuel tank is retained to the frame in a similar manner, using rubber buffers at the front, and a hook-over rubber strap at the rear.

2 The filler cap is recessed into the main body of the tank and is blanked off by a hinged flap that locks by means of a Dzus fastener. This gives the petrol tank a particularly smooth appearance, which conforms with the practice being adopted on other Honda models. The filler cap is removable.

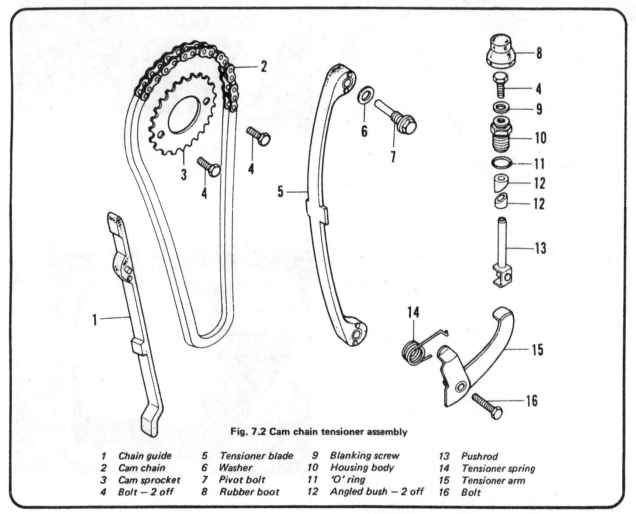

Fig. 7.2 Cam chain tensioner assembly

1 Chain guide
2 Cam chain
3 Cam sprocket
4 Bolt – 2 off
5 Tensioner blade
6 Washer
7 Pivot bolt
8 Rubber boot
9 Blanking screw
10 Housing body
11 'O' ring
12 Angled bush – 2 off
13 Pushrod
14 Tensioner spring
15 Tensioner arm
16 Bolt

Chapter 7: The CB125J model (CB125S-76)

4.1a Clutch operating arm passes through top of side cover

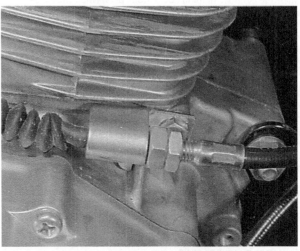
4.1b Cable stop is attached to front crankcase bolt

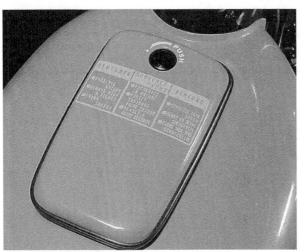
5.2a Open hinged flap ...

5.2b ... to access filler cap

6 Air cleaner assembly: dismantling and cleaning

1 Certain differences occur in the construction of the air cleaner, access to which is gained by removing the right-hand side cover, as in the case of the CB125S model. If the air cleaner cover is unbolted, the air cleaner element within can be withdrawn for cleaning.

2 The wire mesh support is retained by an end cap and is surrounded by a foam rubber sleeve which is the part of the element that will require attention. Wash the element in clean solvent, wring it out and allow it to dry. Impregnate it with clean engine oil, again wring out the excess, and replace the sleeve, making sure it is located correctly and there are no air leaks. The carburettor must be able to breath freely through this sleeve. Finally, replace the air cleaner cover and the right-hand side panel.

7 Exhaust system

1 The exhaust pipe is retained to the cylinder head by a two stud flange fitting arrangement, in which two split collets hold the exhaust pipe firmly in position within the demountable

6.2 Filter element is of foam type, impregnated with oil

Chapter 7: The CB125J model (BC125S-76)

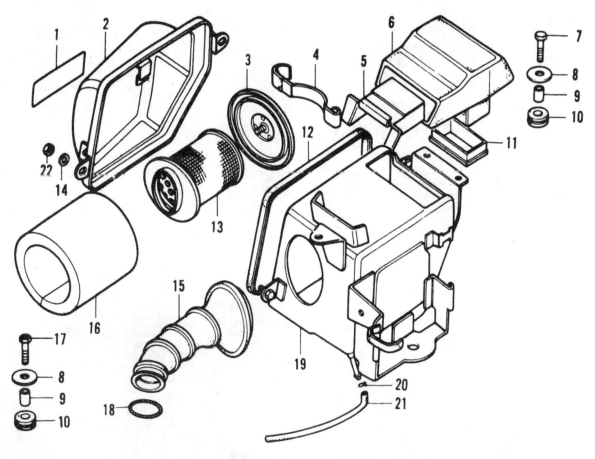

Fig. 7.3. Air cleaner - CB125J model only

1 Air cleaner caution transfer
2 Air cleaner cover
3 Element holder cap
4 Spring clamp
5 Air duct plate A
6 Air duct
7 Bolt - 2 off
8 Washer - 3 off
9 Collar - 3 off
10 Rubber mounting - 3 off
11 Air duct plate B
12 Sealing bead
13 Element support
14 Plain washer - 2 off
15 Carburettor connection
16 Sleeve element
17 Slotted bolt
18 Spring band
19 Air cleaner case
20 Tube clip
21 Tube
22 Nut - 2 off

flange. On the CB125J model, the flange is unfinned.
2 A different design of silencer is used, which is shaped rather like a long, tapering megaphone. A different form of retaining bracket is fitted and there is a bolt-on protective shield to prevent scuffing of the plated finish by the pillion passenger's feet and heat transfer. No attempt should be made to tamper with the silencer or to replace it with one of different design. A more noisy exhaust is not indicative of a gain in performance and in the majority of cases, any such unwarranted change will increase petrol consumption and REDUCE performance, quite apart from rendering the rider liable to prosecution for causing unnecessary noise.

8 Front forks: headlamp mounting and internal construction

1 The front fork assembly fitted to the CB125J model differs in outward appearance in so far as the shrouds over the upper portion of each fork stanchion have been dispensed with, to give the assembly a new, 'slimline' appearance. This has necessitated a revised arrangement for a headlamp mounting, which takes the form of a rod-like structure bent into the form of an inverted letter 'U' which bolts to the upper fork yoke and presses into the lower yoke. Small brackets welded on act as mountings for the headlamp shell, and for the two side-mounted reflectors. A certain amount of anti-vibration insulation is provided by rubber washers at the mounting points.
2 Inwardly, the forks are of similar construction to those fitted to the CB125S models, and use an identical damping arrangement. The most noticeable difference is the use of smaller diameter main fork springs that fit within the stanchions and a different shape of dust cover over the top of the fork sliders. There may also be some minor differences in the number of washers and spacers - refer to the accompanying illustration.
3 When removing the forks from the frame, it is unnecessary to disconnect the brake operating cable from the caliper end of the disc front brake. If the two bolts that retain the left-hand mudguard stay to the fork leg are removed, this will free the upper mounting of the brake caliper. If the lower bolt is then withdrawn, the caliper unit can be lifted away from the forks as a complete unit.

9 Rear suspension units: dismantling

1 The rear suspension units do not have an upper shroud, as

7.2a Silencer is of the long, tapered type

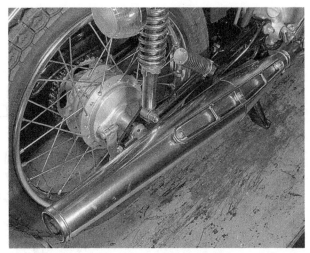
7.2b Note heat shield to protect passenger's feet

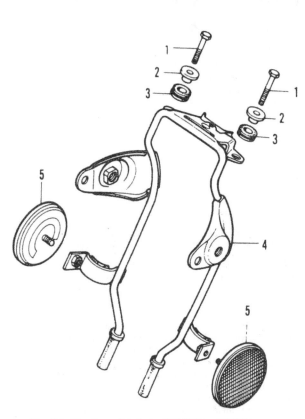

Fig. 7.4. Headlamp bracket - CB125J model only

1 Bolt - 2 off
2 Collar - 2 off
3 Rubber mounting - 2 off
4 Mounting bracket
5 Reflector - 2 off

8.1 Headlamp is mounted on a round section metal framework

3 When changing springs, or even complete suspension units, always do so as a matched pair. In the interests of good road-holding alone, it is important that both units have similar characteristics.

10 Speedometer drive gearbox: location and examination

1 The speedometer drive gearbox is fitted on the right-hand end of the front wheel hub. The gearbox rarely gives trouble unless it is not lubricated regularly. A neglected speedometer drive gearbox will prove stiff to turn and in an extreme case, part of the drive mechanism may shear.
2 The speedometer drive gearbox can be pulled off the end of the front wheel hub after the front wheel has been removed from the forks. Care must be taken to check that the drive engagement dogs align with the dogs on the hub, during reassembly. Check the condition of both sets of dogs to ensure they are not badly worn or broken, and also the condition of the drive pinions.

fitted to the earlier CB125S model. Design changes have made the upper end of each suspension unit detachable, for access to the external spring. It is threaded on to the end of the damper rod. See accompanying illustration.
2 Because a suspension unit is always under tension, it will be necessary to compress the external spring before the upper end can be unscrewed. Honda dealers will have the appropriate tool and it is also possible to purchase an aftermarket tool.

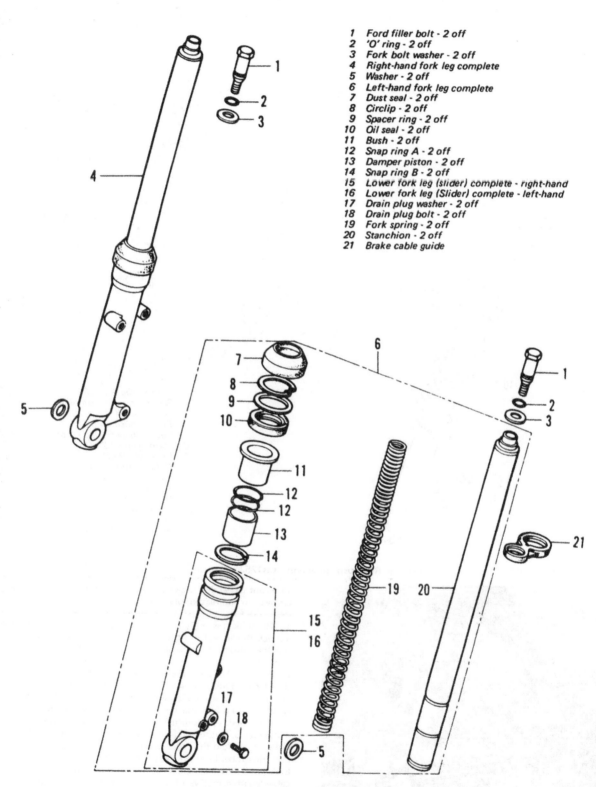

Fig. 7.5. Front forks - CB125J model only

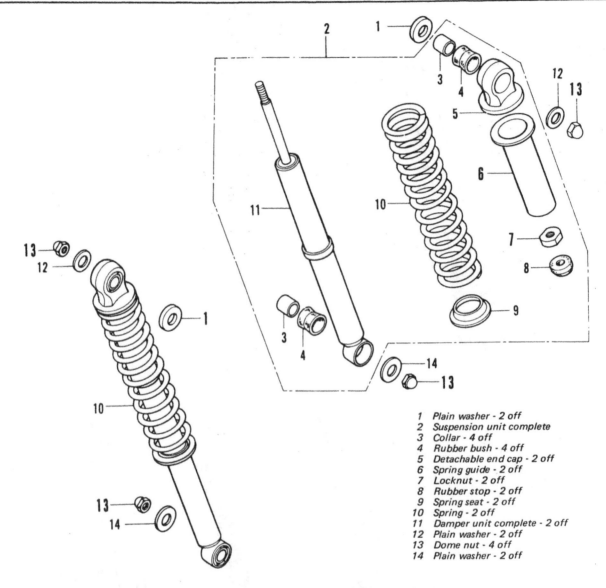

1. Plain washer - 2 off
2. Suspension unit complete
3. Collar - 4 off
4. Rubber bush - 4 off
5. Detachable end cap - 2 off
6. Spring guide - 2 off
7. Locknut - 2 off
8. Rubber stop - 2 off
9. Spring seat - 2 off
10. Spring - 2 off
11. Damper unit complete - 2 off
12. Plain washer - 2 off
13. Dome nut - 4 off
14. Plain washer - 2 off

Fig. 7.6. Rear suspension unit - CB125J model only

9.1 Rear suspension units have exposed springs

11 Front wheel disc brake assembly: examination, dismantling, renovation and reassembly

1 The brake assembly is operated by cable. The cable actuates a quick thread worm within the brake caliper assembly and it is this that presses the pads against the disc when the brake is applied.
2 Each pad has a red line inscribed around its periphery, which denotes the limit of wear. It is usually possible to see the red line on each pad whilst applying the front brake when the assembly is still in situ on the machine with the front wheel in place.
3 If the pads are worn the front wheel must be removed to allow access, and the pads removed as follows: Remove the three bolts which hold the caliper cover in place and detach the cover. Pull the rubber boot off the cable adjuster and loosen the locknut. Screw the adjuster screw as far in as possible. The extra cable length gained will facilitate disconnecting the cable from the quick screw operating arm. Pull the quick screw assembly from position, followed by the thrust plate guide which lies underneath.
4 Screw one of the 6 mm caliper cover screws into one of the threaded holes in the outer disc pad and pull the pad from position. The inner pad is located in the recess in the caliper unit

11.2 Red line around disc brake pads denotes wear limit

11.3 Removal of cover permits access to main adjuster

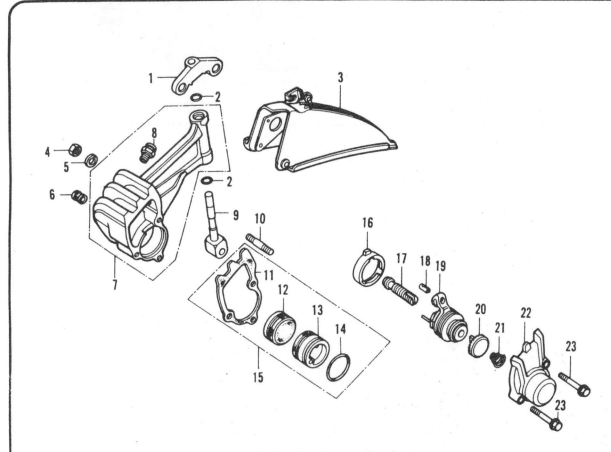

Fig. 7.7. Front brake caliper

1 Caliper mounting	7 Caliper body complete	13 Pad A	19 Operating arm complete
2 'O' ring - 2 off	8 Brake cable bolt	14 'O' ring	20 Adjusting ratchet
3 Disc cover	9 Caliper pivot pin	15 Pad set complete	21 Spring
4 Nut	10 Stud	16 Thrust plate guide	22 Caliper cover
5 Spring washer	11 Caliper gasket	17 Adjuster screw	23 Flange retaining bolt - 3 off
6 Pad grommet	12 Pad B	18 Cable end pin	

by a pin which protrudes through to the right-hand side of the caliper casing. By pressing the head of the locating pin with a suitable tool the pad can be pushed from position.

5 If either pad is worn down to the red 'wear limit' line both pads must be renewed.

6 Before replacing the new pads into the caliper unit make certain that it is absolutely clean. Lightly coat the backing plate of the fixed inner pad with silicone grease suitable for use with brake assemblies. Replace the inner pad in the caliper unit. Refit the front wheel into the forks and tighten the wheel spindle.

7 Place a new 'O' ring on the outer pad and coat the outer circumference of the pad with silicone grease. Replace the pad in the caliper unit making sure that the punch mark on the rear of the pad aligns with the punch mark on the caliper body. Reconnect the brake inner cable with the brake actuating arm and then replace the thrust plate guide. Pull the adjuster ratchet from position on the quick screw assembly and unscrew the adjuster, which lies underneath until it comes up against its stop. Screw in the adjuster screw about ¼ of one turn so that the screw can rotate freely. Replace the ratchet and check that the quick screw assembly functions properly.

8 Fit the quick screw assembly into the caliper unit and replace the cover and gasket which is held by three bolts. Screw the cable adjuster out until any slack in the cable has been removed. Do not over-loosen the adjuster. Now loosen the adjuster a further two turns and tighten the adjuster locknut. Pull the rubber boot down into position over the cable adjuster.

9 No front brake adjustment as such is necessary as the brake will adjust itself automatically when operated. This can be done by operating the handlebar lever 10 times when the self-adjustment will have been accomplished. Check that the front wheel will spin freely and does not drag on the disc pads.

12 Front brake disc: removal, examination and replacement

1 After extended service the front brake disc will wear down to a point where it cannot function efficiently. The thickness of the disc can be measured at various points around its circumference with the use of a micrometer. If the disc wears below 0.0157 in.

11.8 Make fine adjustments with this external cable adjuster

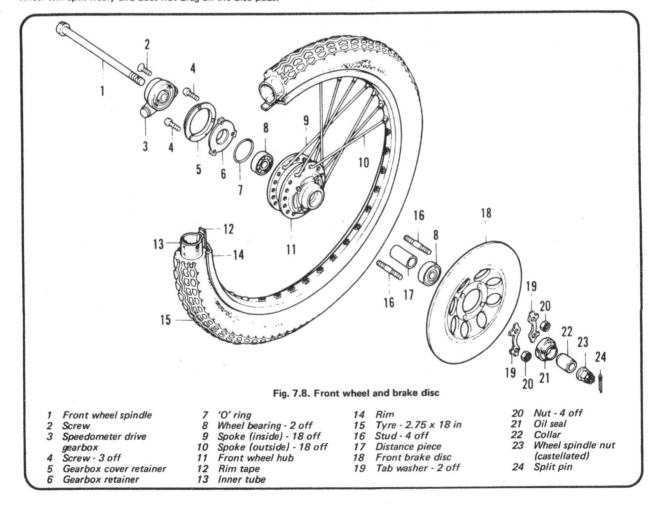

Fig. 7.8. Front wheel and brake disc

1 Front wheel spindle	7 'O' ring	14 Rim
2 Screw	8 Wheel bearing - 2 off	15 Tyre - 2.75 x 18 in
3 Speedometer drive gearbox	9 Spoke (inside) - 18 off	16 Stud - 4 off
4 Screw - 3 off	10 Spoke (outside) - 18 off	17 Distance piece
5 Gearbox cover retainer	11 Front wheel hub	18 Front brake disc
6 Gearbox retainer	12 Rim tape	19 Tab washer - 2 off
	13 Inner tube	20 Nut - 4 off
		21 Oil seal
		22 Collar
		23 Wheel spindle nut (castellated)
		24 Split pin

(4 mm) it should be replaced. Check the disc for warpage by spinning the wheel in the forks. Warpage of more than 0.0079 in. (2.0 mm) measured at the inner radius of the disc will reduce the efficiency of the brake and will cause juddering and premature pad wear.

2 The disc is retained on the wheel hub by four studs and nuts and can be removed after the front wheel has been detached from the front forks. The locknuts are secured in pairs by common tab washers which must be knocked down before the nuts are removed.

13 Wheel bearings: examination and replacement

1 To gain access to the front wheel bearings on the disc brake models, it is first necessary to remove the speedometer drive gearbox, which will pull off, the gearbox retainer and the retainer cover. The retainer and cover are retained by three cross head screws. Prise the oil seal from position in the left-hand end of the wheel hub.

2 The bearings can then be driven out as described in Section 4 of Chapter 5 and the remainder of the procedure in that Section followed.

14 Front wheel disc brake: adjustment

1 As mentioned previously, the front wheel disc brake is of the self-adjusting type and will automatically take up excess play as the disc pads wear. It is imperative that the pads are not permitted to wear beyond the red line inscribed around their periphery, a point that is easily overlooked as the brake does not require the regular attention of a drum brake.

15 Other minor differences - CB125J model

1 When compared to the earlier CB125S model, other minor differences will be noted here and there on the CB125J model. These are, however, only of a superficial nature, brought about by changes in design of other components with which they are associated and which have been mentioned previously, or by restyling. They should not in anyway necessitate revision of the dismantling and reassembly procedure sequences described in earlier Chapters.

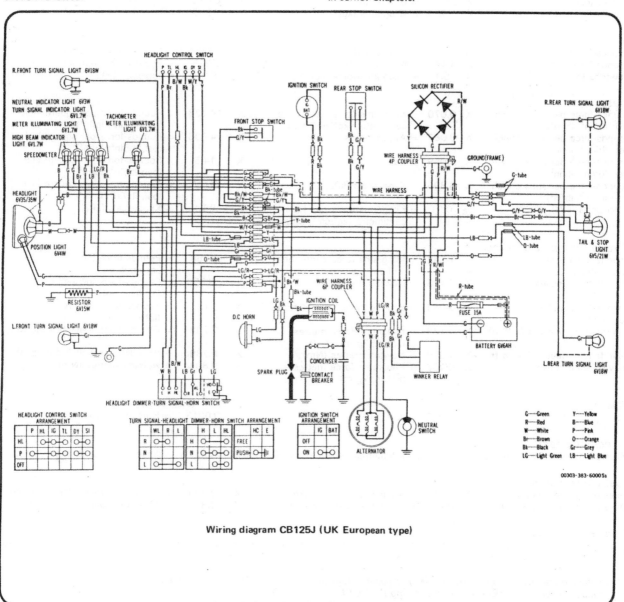

Wiring diagram CB125J (UK European type)

Conversion factors

Length (distance)
Inches (in)	X	25.4 = Millimetres (mm)	X	0.0394	= Inches (in)
Feet (ft)	X	0.305 = Metres (m)	X	3.281	= Feet (ft)
Miles	X	1.609 = Kilometres (km)	X	0.621	= Miles

Volume (capacity)
Cubic inches (cu in; in^3)	X	16.387 = Cubic centimetres (cc; cm^3)	X	0.061	= Cubic inches (cu in; in^3)
Imperial pints (Imp pt)	X	0.568 = Litres (l)	X	1.76	= Imperial pints (Imp pt)
Imperial quarts (Imp qt)	X	1.137 = Litres (l)	X	0.88	= Imperial quarts (Imp qt)
Imperial quarts (Imp qt)	X	1.201 = US quarts (US qt)	X	0.833	= Imperial quarts (Imp qt)
US quarts (US qt)	X	0.946 = Litres (l)	X	1.057	= US quarts (US qt)
Imperial gallons (Imp gal)	X	4.546 = Litres (l)	X	0.22	= Imperial gallons (Imp gal)
Imperial gallons (Imp gal)	X	1.201 = US gallons (US gal)	X	0.833	= Imperial gallons (Imp gal)
US gallons (US gal)	X	3.785 = Litres (l)	X	0.264	= US gallons (US gal)

Mass (weight)
Ounces (oz)	X	28.35 = Grams (g)	X	0.035	= Ounces (oz)
Pounds (lb)	X	0.454 = Kilograms (kg)	X	2.205	= Pounds (lb)

Force
Ounces-force (ozf; oz)	X	0.278 = Newtons (N)	X	3.6	= Ounces-force (ozf; oz)
Pounds-force (lbf; lb)	X	4.448 = Newtons (N)	X	0.225	= Pounds-force (lbf; lb)
Newtons (N)	X	0.1 = Kilograms-force (kgf; kg)	X	9.81	= Newtons (N)

Pressure
Pounds-force per square inch (psi; lbf/in^2; lb/in^2)	X	0.070 = Kilograms-force per square centimetre (kgf/cm^2; kg/cm^2)	X	14.223	= Pounds-force per square inch (psi; lbf/in^2; lb/in^2)
Pounds-force per square inch (psi; lbf/in^2; lb/in^2)	X	0.068 = Atmospheres (atm)	X	14.696	= Pounds-force per square inch (psi; lbf/in^2; lb/in^2)
Pounds-force per square inch (psi; lbf/in^2; lb/in^2)	X	0.069 = Bars	X	14.5	= Pounds-force per square inch (psi; lbf/in^2; lb/in^2)
Pounds-force per square inch (psi; lbf/in^2; lb/in^2)	X	6.895 = Kilopascals (kPa)	X	0.145	= Pounds-force per square inch (psi; lbf/in^2; lb/in^2)
Kilopascals (kPa)	X	0.01 = Kilograms-force per square centimetre (kgf/cm^2; kg/cm^2)	X	98.1	= Kilopascals (kPa)

Torque (moment of force)
Pounds-force inches (lbf in; lb in)	X	1.152 = Kilograms-force centimetre (kgf cm; kg cm)	X	0.868	= Pounds-force inches (lbf in; lb in)
Pounds-force inches (lbf in; lb in)	X	0.113 = Newton metres (Nm)	X	8.85	= Pounds-force inches (lbf in; lb in)
Pounds-force inches (lbf in; lb in)	X	0.083 = Pounds-force feet (lbf ft; lb ft)	X	12	= Pounds-force inches (lbf in; lb in)
Pounds-force feet (lbf ft; lb ft)	X	0.138 = Kilograms-force metres (kgf m; kg m)	X	7.233	= Pounds-force feet (lbf ft; lb ft)
Pounds-force feet (lbf ft; lb ft)	X	1.356 = Newton metres (Nm)	X	0.738	= Pounds-force feet (lbf ft; lb ft)
Newton metres (Nm)	X	0.102 = Kilograms-force metres (kgf m; kg m)	X	9.804	= Newton metres (Nm)

Power
Horsepower (hp)	X	745.7 = Watts (W)	X	0.0013	= Horsepower (hp)

Velocity (speed)
Miles per hour (miles/hr; mph)	X	1.609 = Kilometres per hour (km/hr; kph)	X	0.621	= Miles per hour (miles/hr; mph)

Fuel consumption*
Miles per gallon, Imperial (mpg)	X	0.354 = Kilometres per litre (km/l)	X	2.825	= Miles per gallon, Imperial (mpg)
Miles per gallon, US (mpg)	X	0.425 = Kilometres per litre (km/l)	X	2.352	= Miles per gallon, US (mpg)

Temperature
Degrees Fahrenheit = (°C x 1.8) + 32   Degrees Celsius (Degrees Centigrade; °C) = (°F - 32) x 0.56

*It is common practice to convert from miles per gallon (mpg) to litres/100 kilometres (l/100km), where mpg (Imperial) x l/100 km = 282 and mpg (US) x l/100 km = 235

Index

A

Air cleaner - 44, 96
Advance/retard mechanism - 50
Alternator:
   output check - 79
   removal - 17
   replacement - 35

B

Balancing, front wheel - 75
Battery:
   charging - 80, 81
   maintenance - 80
   specifications - 79
Big-end - 21
Brakes:
   adjustment - 75, 103
   fault diagnosis - 78
   front - examination, dismantling & reassembly - 70
   front disc assembly - 100, 102, 103
   rear - examination, dismantling & reassembly - 72
   specifications - 69
Bulb replacement:
   headlamp - 81
   indicators - 82
   stop lamp - 82
   tail lamp - 82
Bulb specifications - 69

C

Cam followers (rockers), inspection - 21
Camshaft:
   inspection - 21
   replacement - 35
   removal - 15
Camshaft chain:
   guide, replacement - 35
   inspection - 21
   replacement - 35
   removal - 17
   sprockets, inspection - 21
   tensioner - 35, 37
Capacities - 7
Carburettor:
   adjusting - 43
   dismantling, inspection & reassembly - 42
   removal - 42
   specifications - 41, 92
Centre stand - 66
Cleaning - 67
Clutch:
   fault diagnosis - 40
   inspection - 27
   operation (CB125J model) - 95
   reassembly and replacement - 32
   removal and dismantling - 19
   specifications - 11
Compression testing - 38
Condenser - 50
Contact breaker - 49
Crankcases:
   joining - 32
   separating - 21
Crankshaft:
   inspection - 21
   replacement - 24
   removal - 21
Cylinder barrel:
   inspection - 21
   replacement - 35
   removal - 15
Cylinder head:
   decarbonising - 23
   dismantling and inspection - 23, 94
   replacement - 35
   removal - 15

D

Dimensions - 4, 91
Dualseat - 67

E

Electrical system, specifications - 79
Emergency switch - 83
Engine:
   dismantling - 14, 15, 17, 19 and 21
   fault diagnosis - 39
   reassembly - 29, 32, 35, 37 and 38
   specifications - 9, 10, 11, 91 and 92
Engine cover - 35
Engine/gearbox unit:
   replacement - 38
   removal - 11
Exhaust system - 45, 96

F

Final drive chain:
   adjustment - 75
   examination and lubrication - 75
   final drive sprocket - 17
Flasher unit - 82
Footrest bar - 66
Frame:
   examination and renovation - 61
   fault diagnosis - 68
   specifications - 54
Front brake - 70
Front forks:
   description (CB125J model) - 97
   fault diagnosis - 68
   reassembly and replacement - 58
   removal, dismantling and inspection - 56
   specifications - 54, 92
Front wheel:
   examination and removal - 69
   replacement - 72
Front wheel bearings - 72
Fuel system:
   fault diagnosis - 48
   specifications - 41
Fuel tap - 41
Fuel tank - 41
Fuses - 79 and 83

Index

G

Gearbox:
    components - examination and renovation - 25 and 27
    dismantling - 25
    fault diagnosis - 40
    reassembly - 29
    specifications - 11, 91
Gearbox bearings - 25
Gearbox/engine:
    replacement - 38
    removal - 11
Gearbox oil seals - 25
Gear cluster:
    replacement - 29
    removal - 21
Gear selector mechanism:
    dismantling - 19
    inspection - 29
    reassembly - 29 and 32

H

Headlamp:
    adjustment - 81
    bulb replacement - 81
    mounting (CB125J model) - 97
    dipswitch and pilot light - 83
    switch - 83
Horn - 83

I

Ignition coil - 49
Ignition cut-out - 83
Ignition switch - 83
Ignition system:
    fault diagnosis - 53
    specifications - 49
Ignition timing - 49
Indicators - 82

K

Kickstarter - 29

M

Main bearings - 21
Maintenance, routine - 6
Modifications - 5

O

Oil filter - 19, 32 and 47
Oil pump:
    general - 47
    replacement - 32
    removal - 19
Oil seals - 21

P

Petrol feed pipe - 42
Petrol tank:
    badges, removal and replacement - 67
    description (CB125J model) - 95
    removal and replacement - 41
Petrol tap - 41
Piston:
    inspection - 21
    replacement - 35
    removal - 15 and 17
Piston rings - 21
Primary drive - 27 and 32
Prop stand - 66

R

Rear brake - 72
Rear suspension units - 62, 65, 97
Rear wheel:
    examination and removal - 72
    replacement - 75
Rear wheel bearings - 72
Rear wheel sprocket - 72
Rectifier - 80
Resistor - 81
Rockers (cam followers):
    inspection - 21
    removal - 21
Routine maintenance - 6
Running-in - 38

S

Seat - 67
Selenium rectifier - 80
Side covers - 65, 95
Silencer - 45
Spark plug - 53
Spark plug chart (colour) - 51
Speedometer:
    cable - examination and renovation - 66
    drive gearbox - 98
    head - removal and replacement - 66
    indicator lamps - bulb replacement - 82
Sprockets:
    camshaft - 21
    final drive - 17
    rear wheel - 72
Steering head races - 58
Steering lock - 67
Stop lamp:
    bulb replacement - 82
    switch - 82
Suspension:
    fault diagnosis - 68
    specifications - 54
Swinging arm - 62

T

Tachometer:
    cable - removal and replacement - 66
    head - removal and replacement - 66
    indicator bulb - replacement - 82
Tail lamp, bulb replacement - 82
Throttle cable - 44
Tyres:
    changing sequence (colour) - 77
    fault diagnosis - 78
    pressures - 69
    replacement - 76
    removal - 76
    security bolts - 76
    sizes - 69, 92

V

Valves:
    clearances (tappets) - 37
    inspection - 23
    grinding-in - 23
    guides - removal and replacement - 23
    removal and replacement - 23
    timing - 35 and 37

W

Weight - 4
Wheel balance - 75
Wheel bearings - 72, 103
Wheel (front) - 69 and 72
Wheel (rear) - 72 and 75
Width - 4
Wiring diagrams - 84, 85, 86, 87, 88 and 89
Wiring, layout and examination - 83

English/American terminology

Because this book has been written in England, British English component names, phrases and spellings have been used throughout. American English usage is quite often different and whereas normally no confusion should occur, a list of equivalent terminology is given below.

English	American	English	American
Air filter	Air cleaner	Number plate	License plate
Alignment (headlamp)	Aim	Output or layshaft	Countershaft
Allen screw/key	Socket screw/wrench	Panniers	Side cases
Anticlockwise	Counterclockwise	Paraffin	Kerosene
Bottom/top gear	Low/high gear	Petrol	Gasoline
Bottom/top yoke	Bottom/top triple clamp	Petrol/fuel tank	Gas tank
Bush	Bushing	Pinking	Pinging
Carburettor	Carburetor	Rear suspension unit	Rear shock absorber
Catch	Latch	Rocker cover	Valve cover
Circlip	Snap ring	Selector	Shifter
Clutch drum	Clutch housing	Self-locking pliers	Vise-grips
Dip switch	Dimmer switch	Side or parking lamp	Parking or auxiliary light
Disulphide	Disulfide	Side or prop stand	Kick stand
Dynamo	DC generator	Silencer	Muffler
Earth	Ground	Spanner	Wrench
End float	End play	Split pin	Cotter pin
Engineer's blue	Machinist's dye	Stanchion	Tube
Exhaust pipe	Header	Sulphuric	Sulfuric
Fault diagnosis	Trouble shooting	Sump	Oil pan
Float chamber	Float bowl	Swinging arm	Swingarm
Footrest	Footpeg	Tab washer	Lock washer
Fuel/petrol tap	Petcock	Top box	Trunk
Gaiter	Boot	Torch	Flashlight
Gearbox	Transmission	Two/four stroke	Two/four cycle
Gearchange	Shift	Tyre	Tire
Gudgeon pin	Wrist/piston pin	Valve collar	Valve retainer
Indicator	Turn signal	Valve collets	Valve cotters
Inlet	Intake	Vice	Vise
Input shaft or mainshaft	Mainshaft	Wheel spindle	Axle
Kickstart	Kickstarter	White spirit	Stoddard solvent
Lower leg	Slider	Windscreen	Windshield
Mudguard	Fender		

Printed and bound by CPI Group (UK) Ltd, Croydon, CR0 4YY
09/06/2025

14685667-0003